# Introduction

**Kakuro** (also known as *"Cross Sums"*) is a mathematical equivalent of crosswords. The game board resembles a crossword grid with both filled and blank squares. Some of the filled, black squares have a diagonal line with numbers in each half, which serve as the puzzle's hints. The number in the upper half of the square gives a clue for the horizontal (across) sum of the numbers, while the number in the lower half is a clue for the vertical (down) sum.

The aim of Kakuro is to place numbers from 1 to 9 into the blank squares so that they add up to the clues given. Importantly, no number may be used more than once in any given sum. For instance, if the clue is 6, acceptable combinations include 1 and 5 or 2 and 4, but not 3 and 3, as that repeats the number.

Copyright © 2024 Julian Blake
All Rights Reserved.

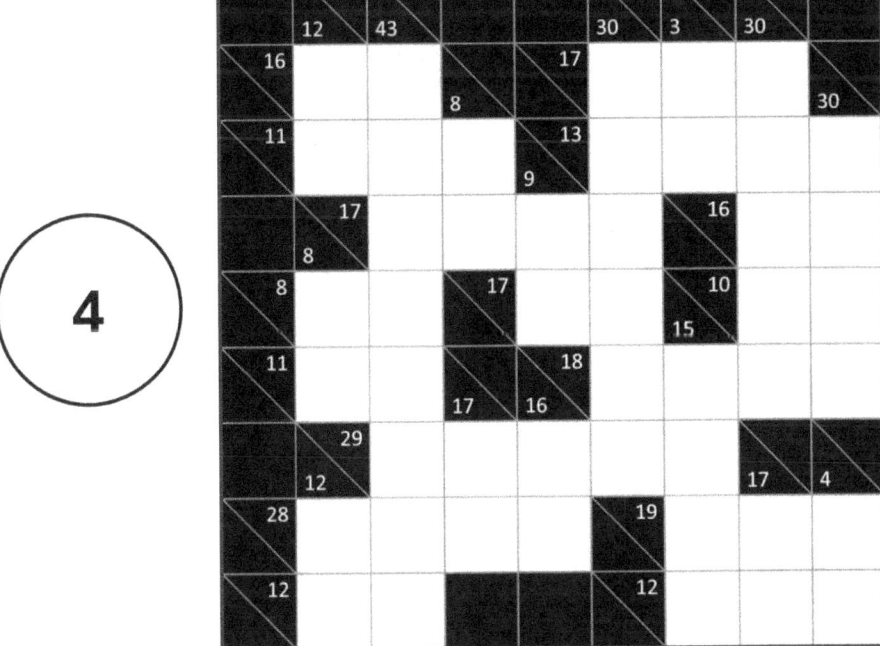

**7**

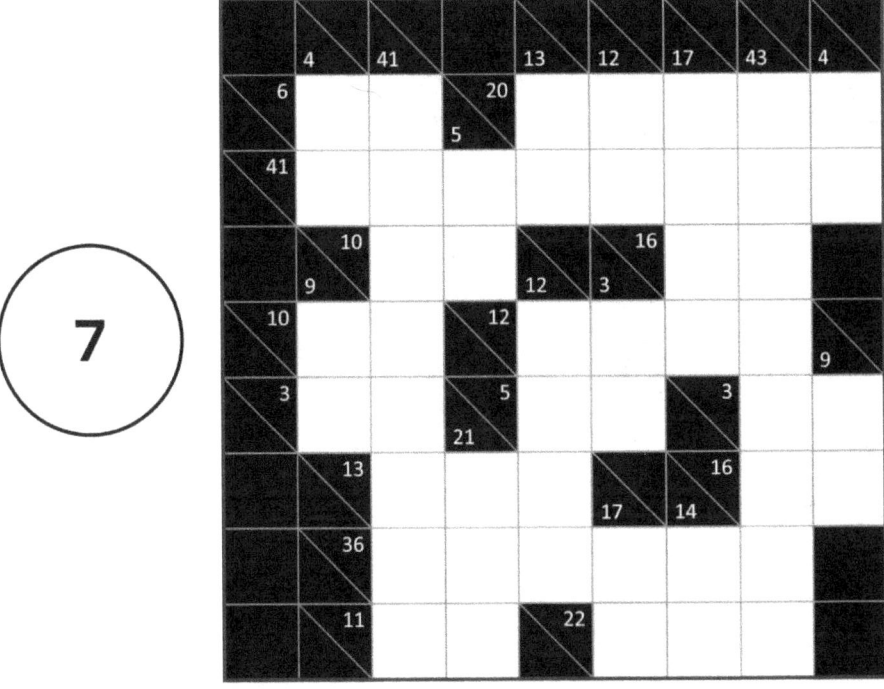

**8**

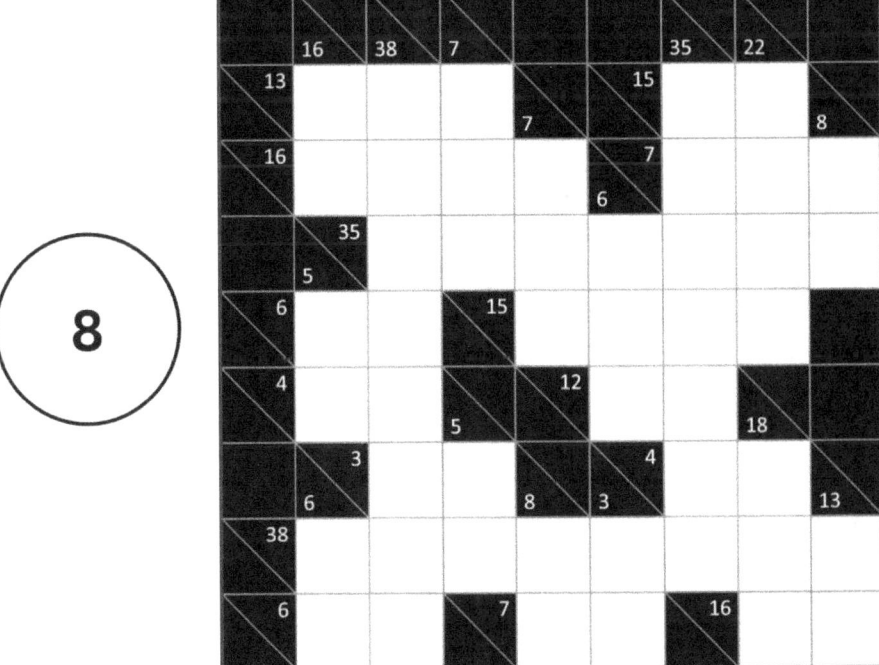

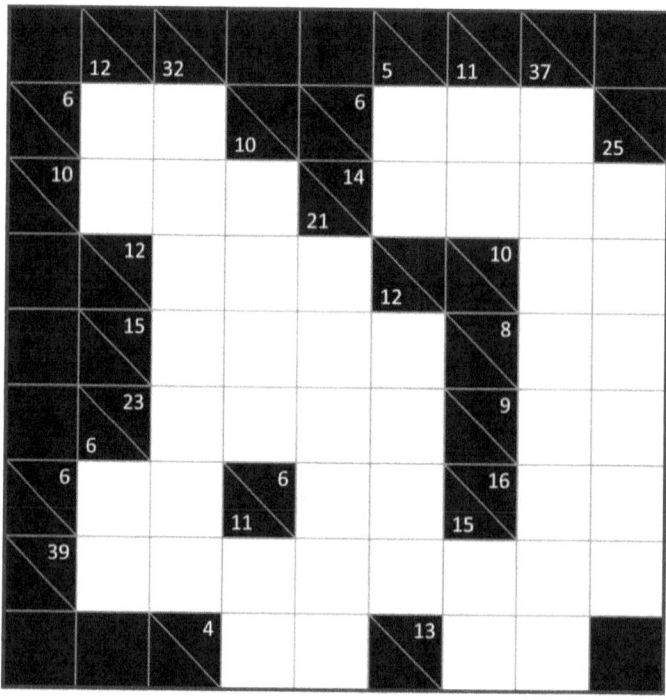

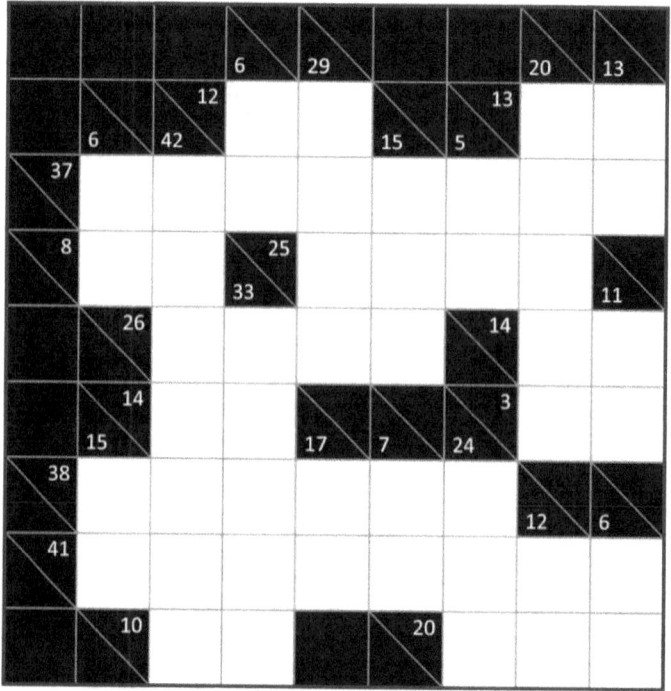

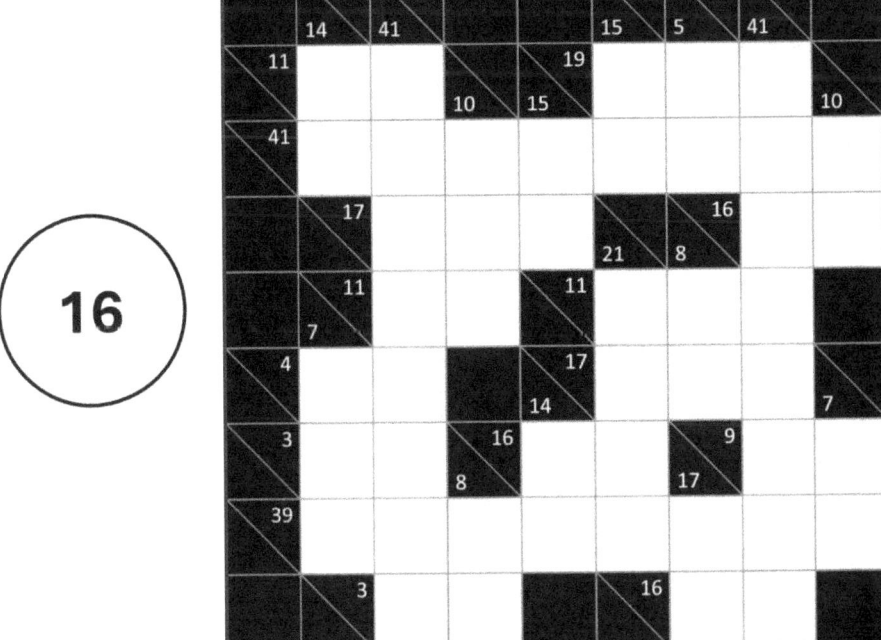

**15**

**16**

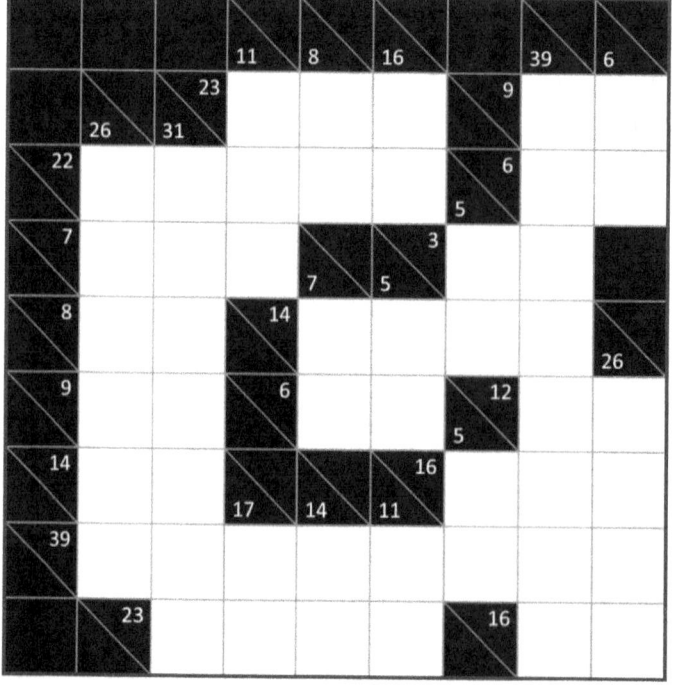

# 19

# 20

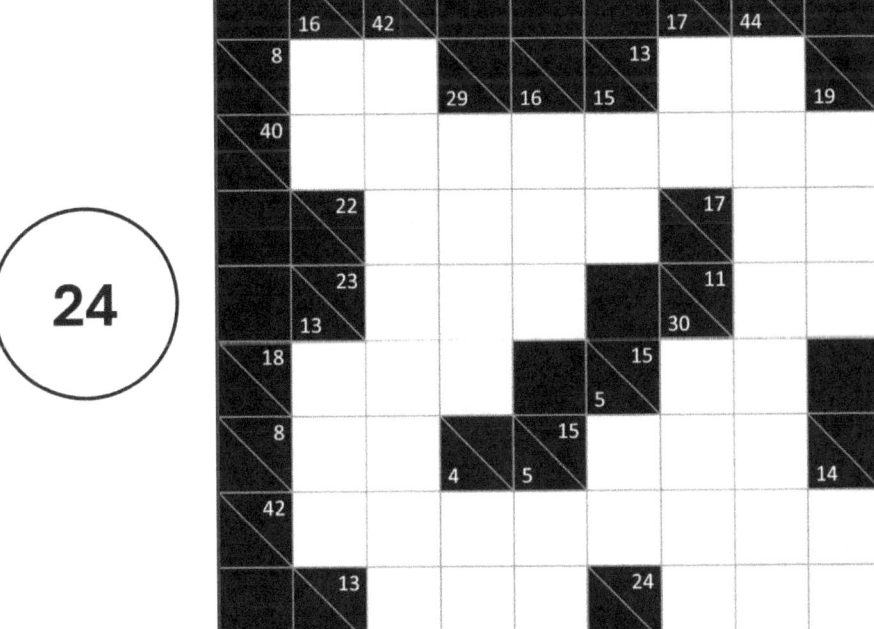

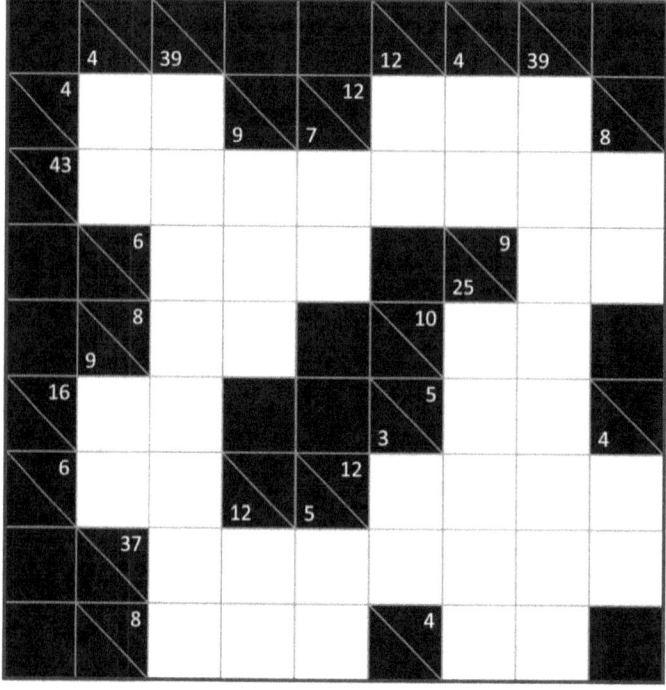

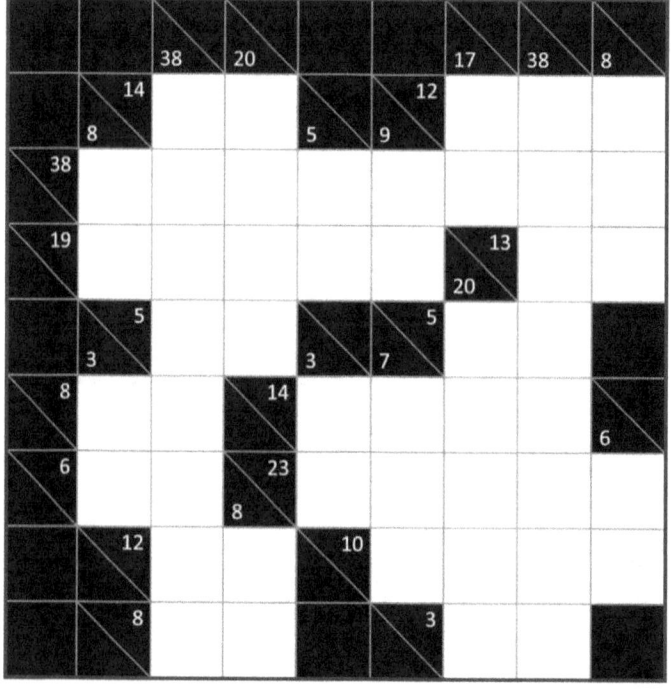

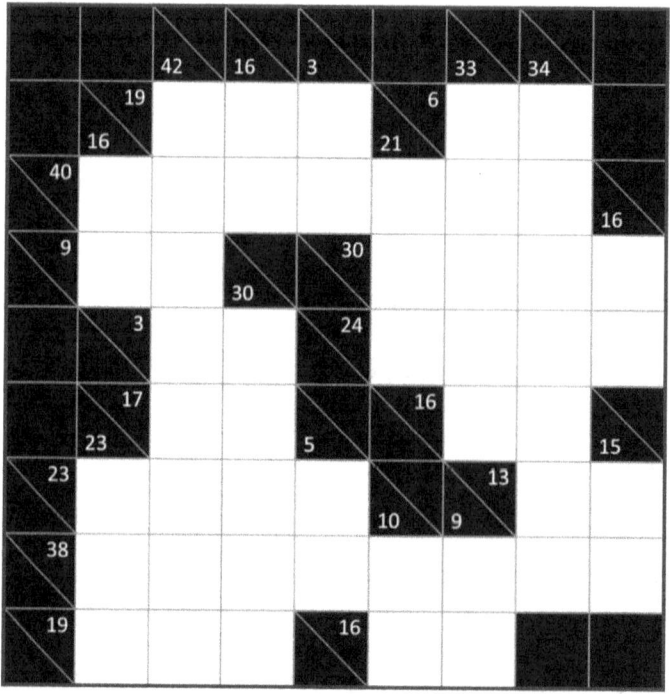

35

36

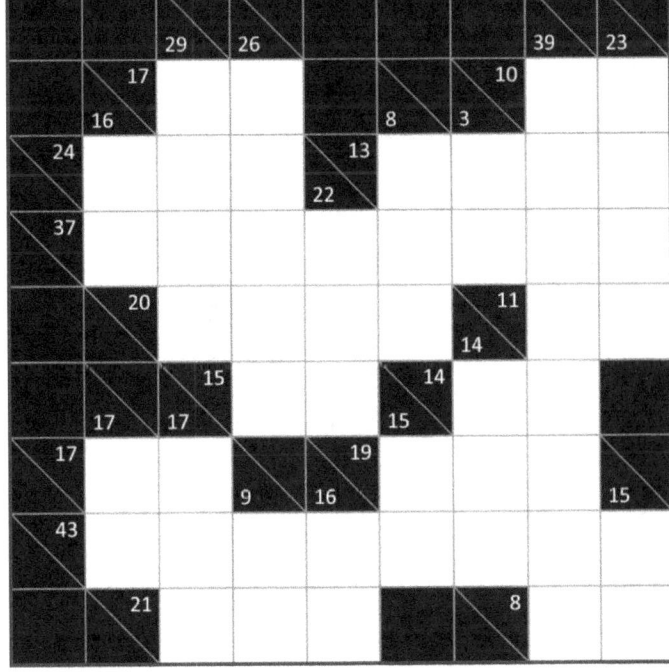

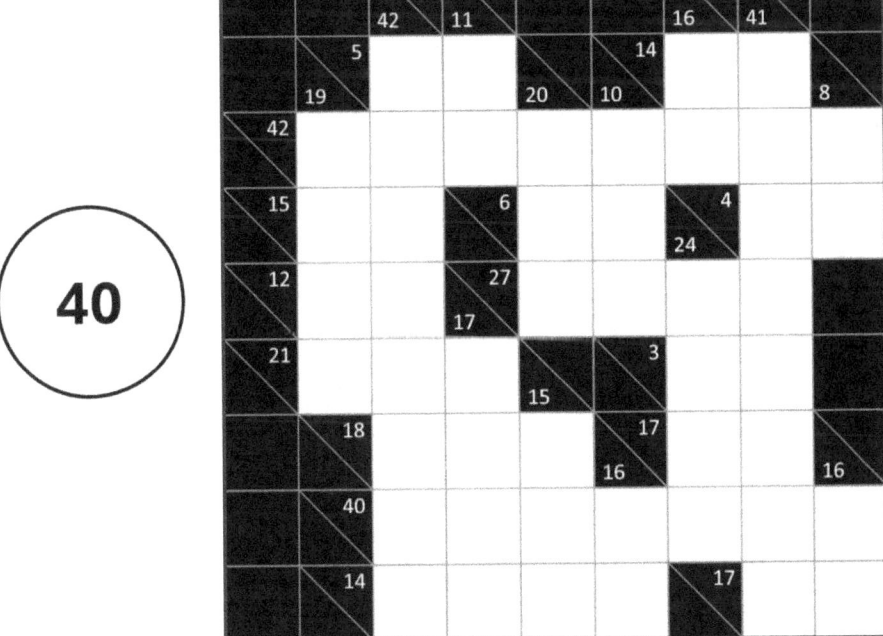

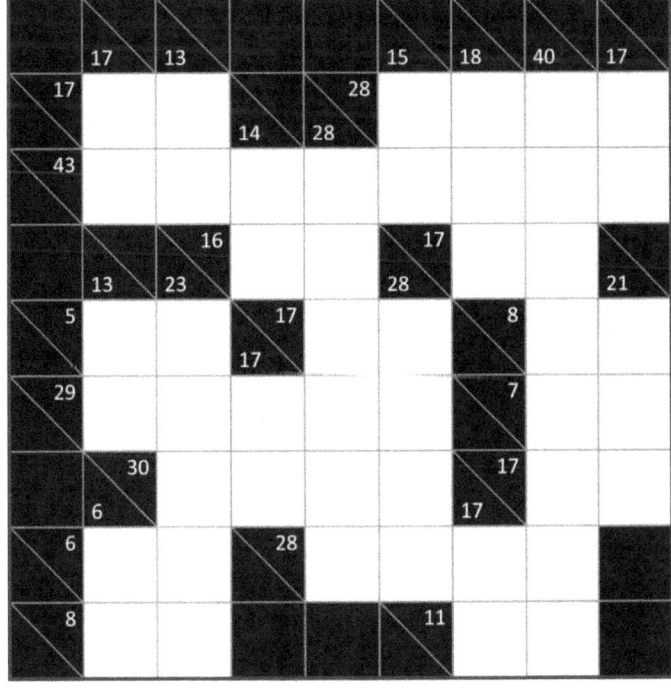

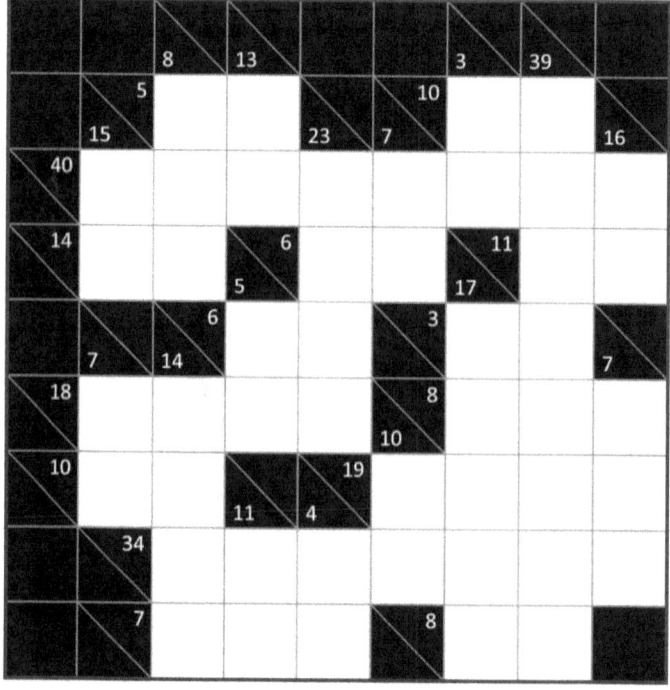

**47**

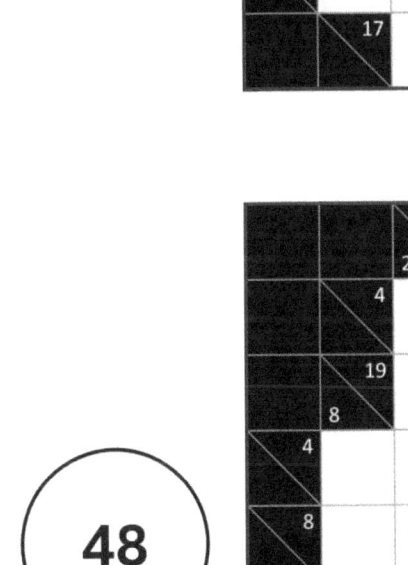

**48**

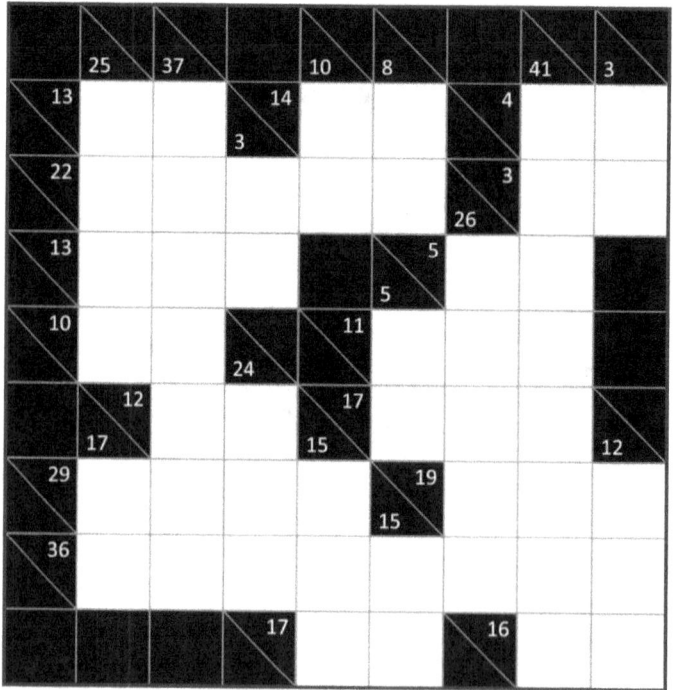

51

52

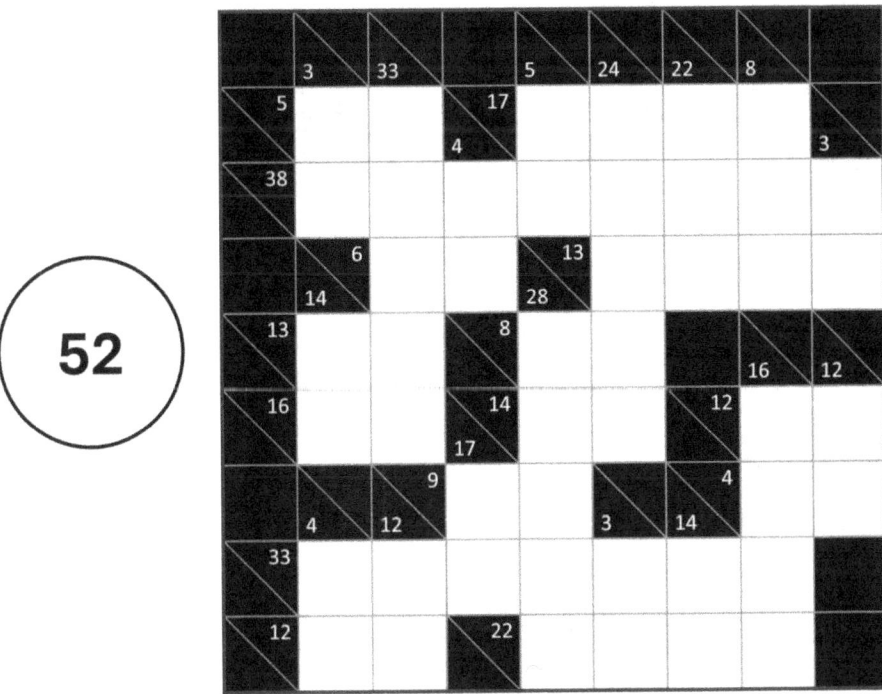

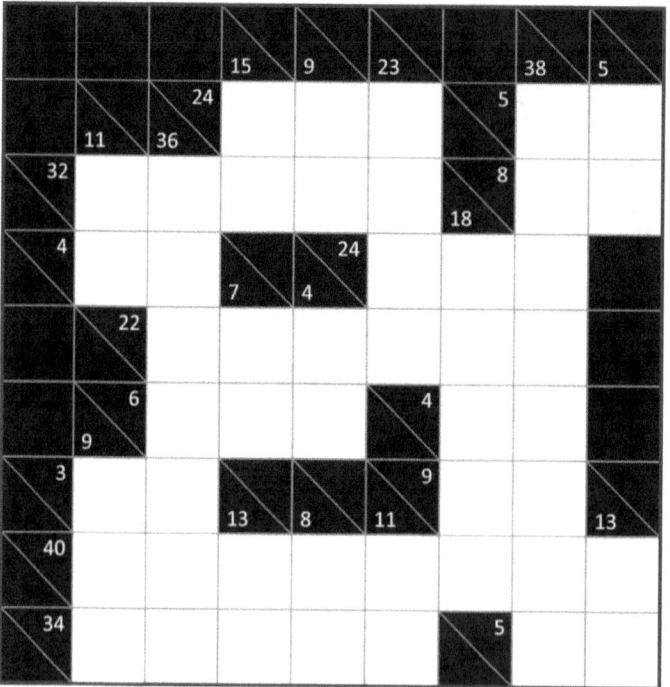

**55**

**56**

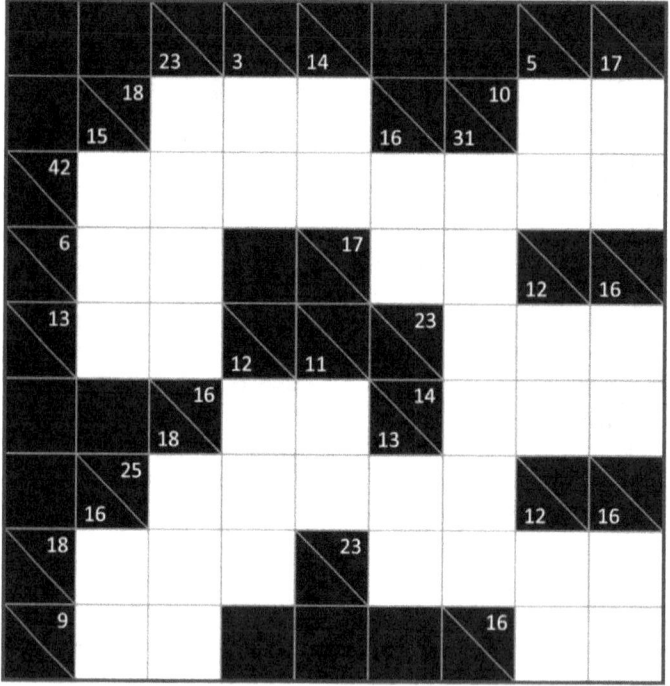

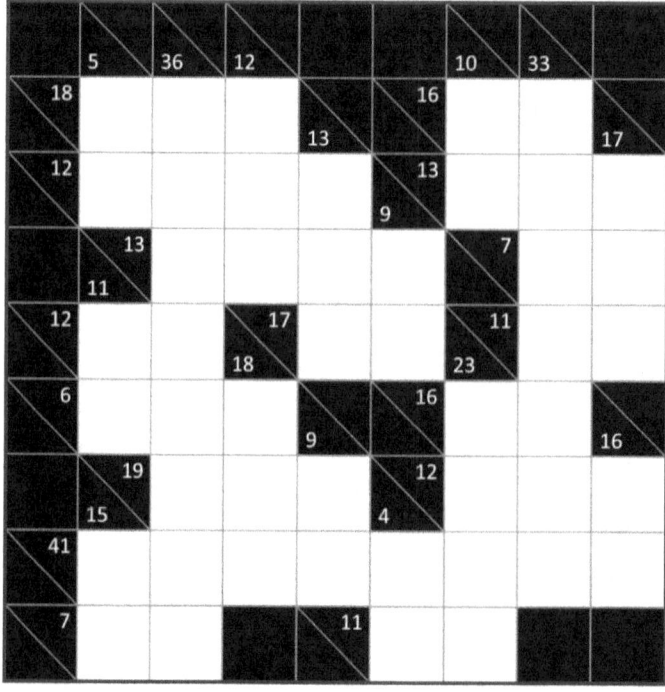

59

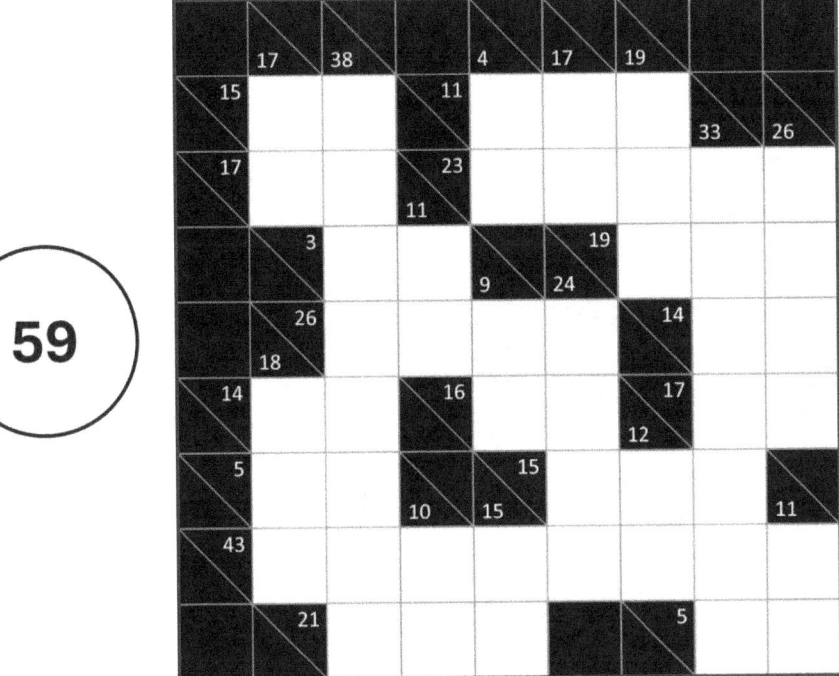

60

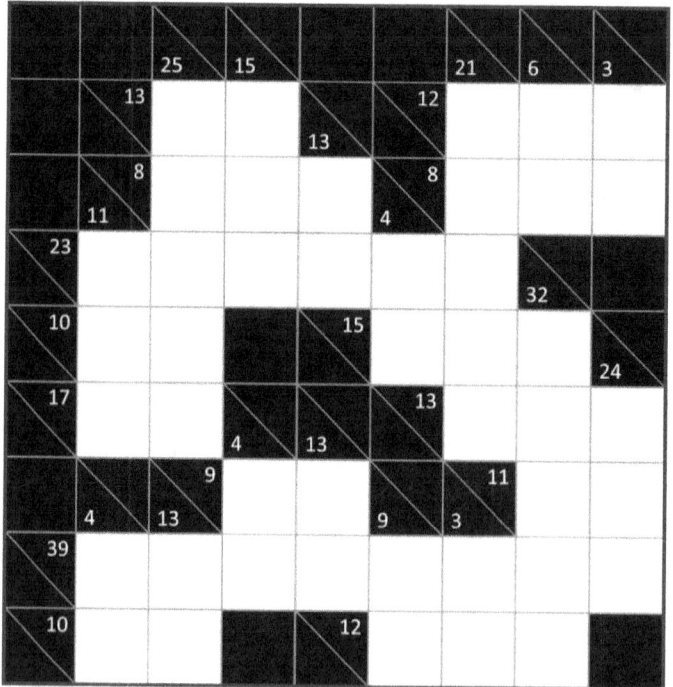

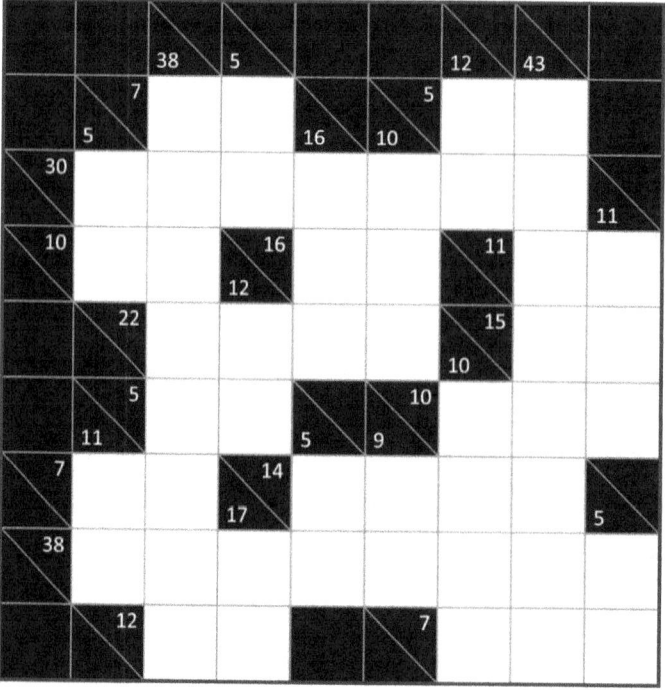

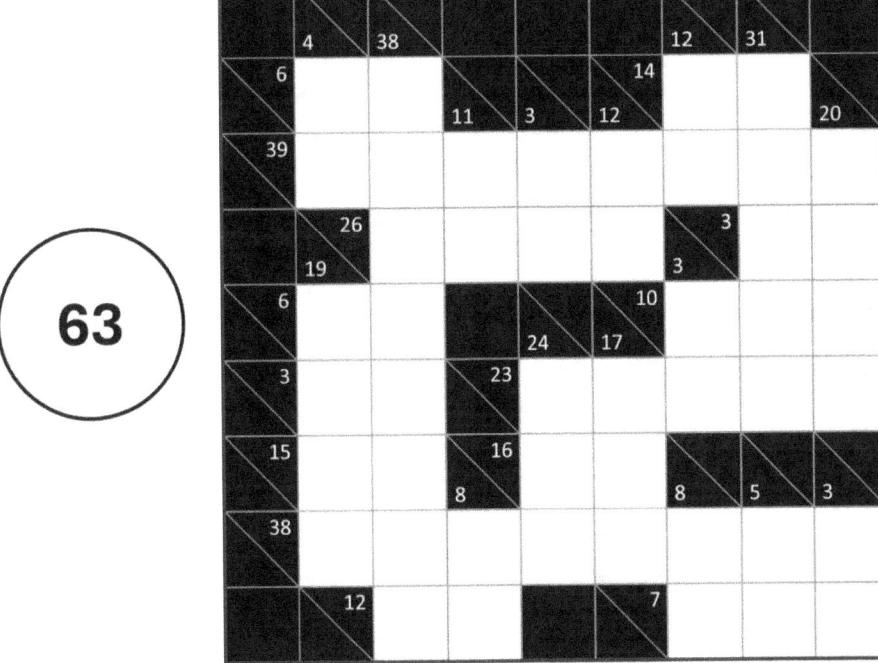

**67**

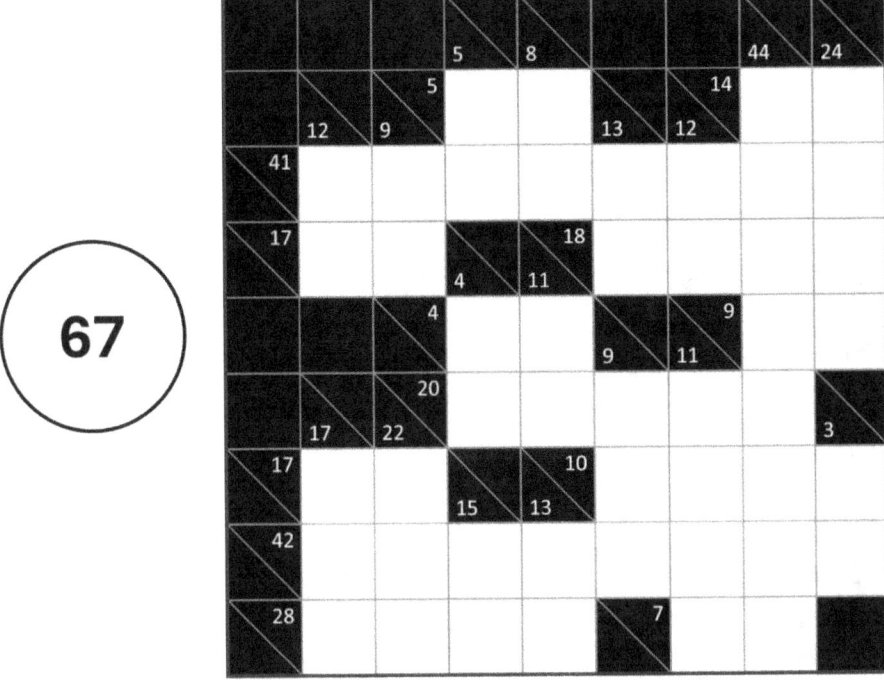

**68**

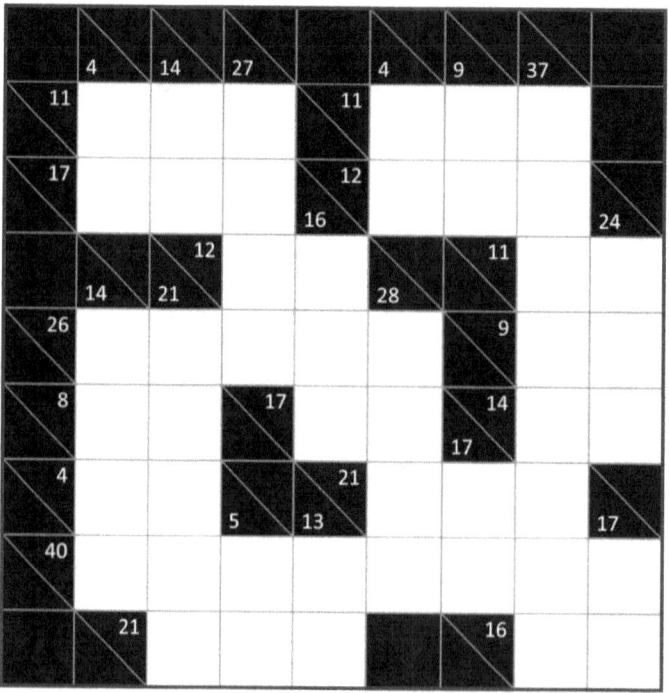

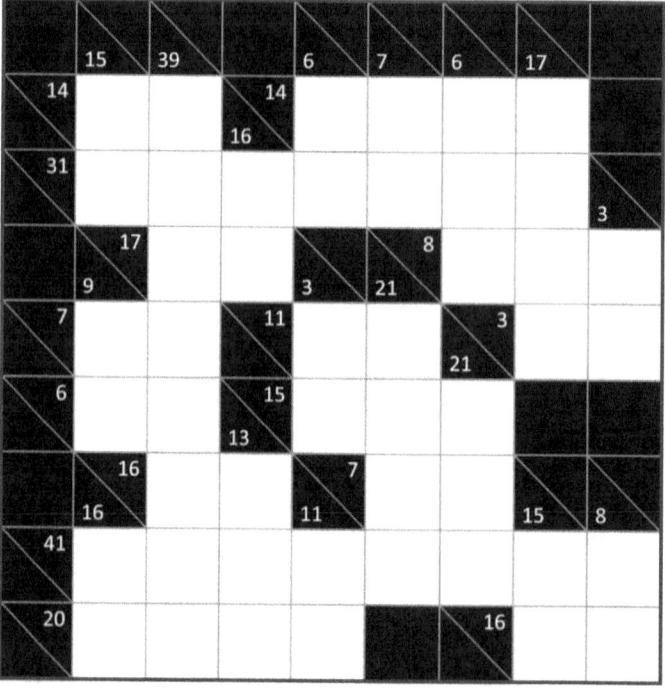

# 71

# 72

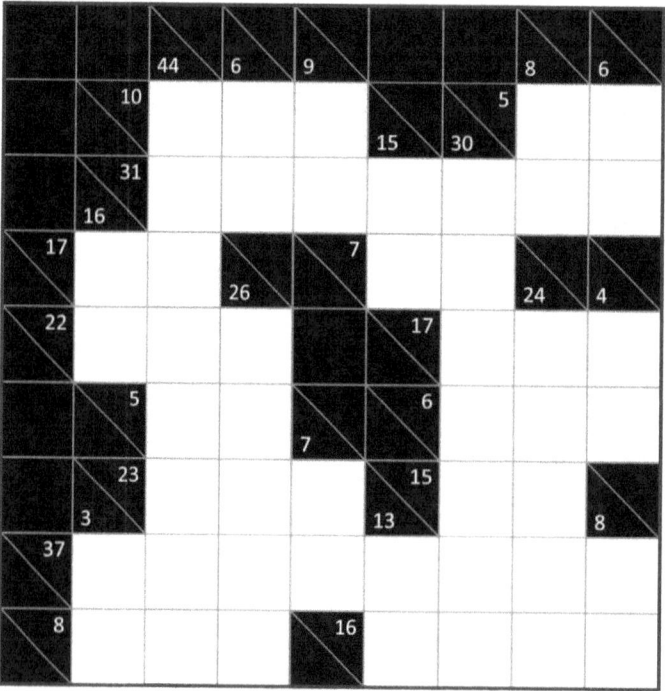

**75**

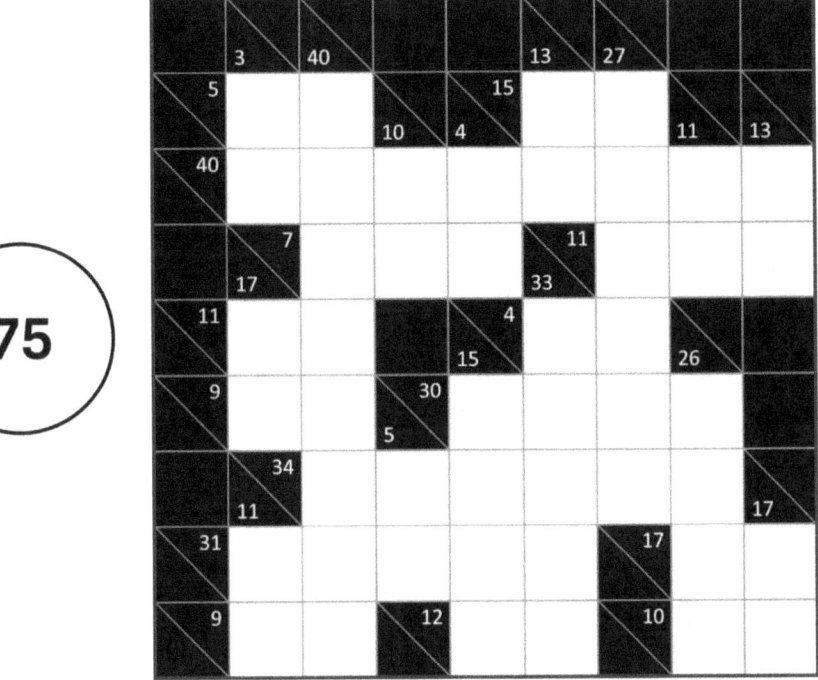

**76**

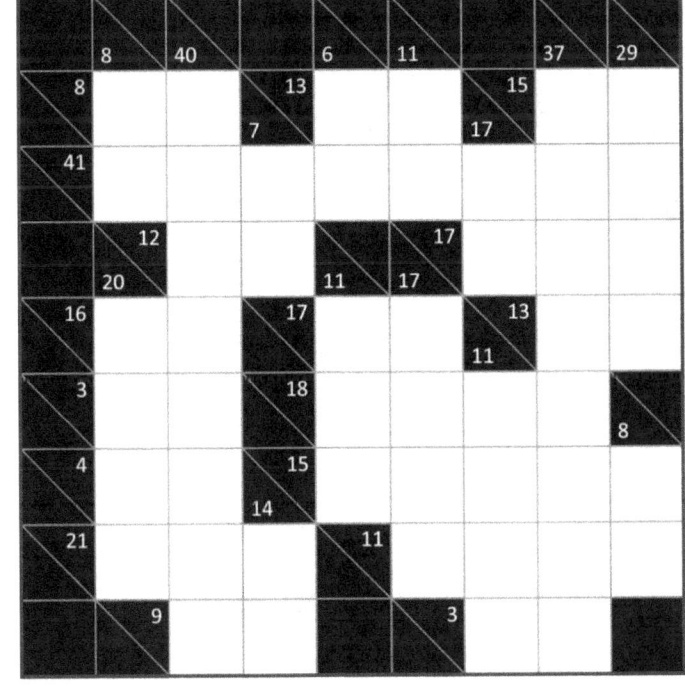

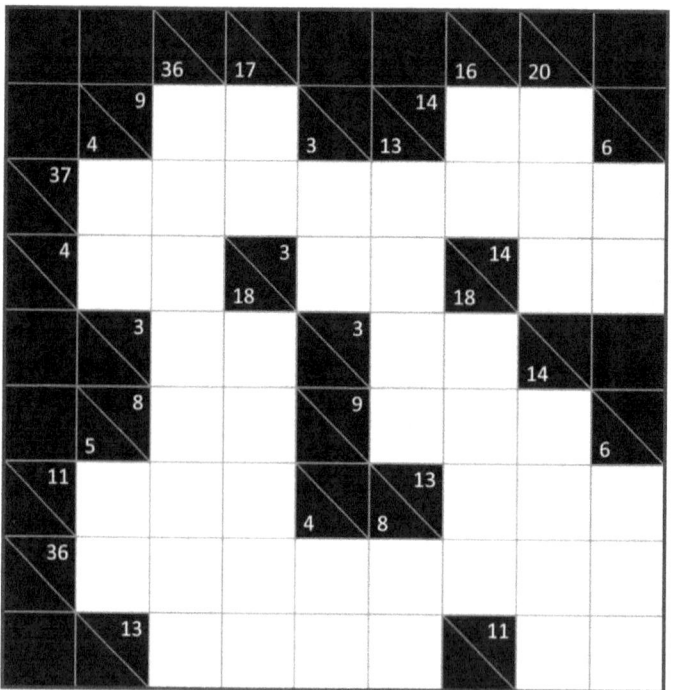

**79**

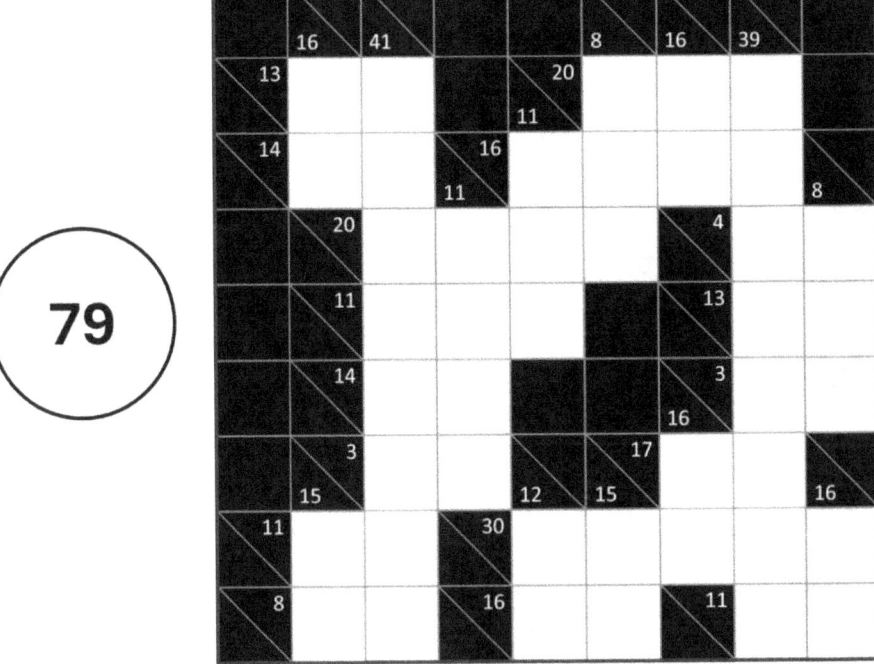

**80**

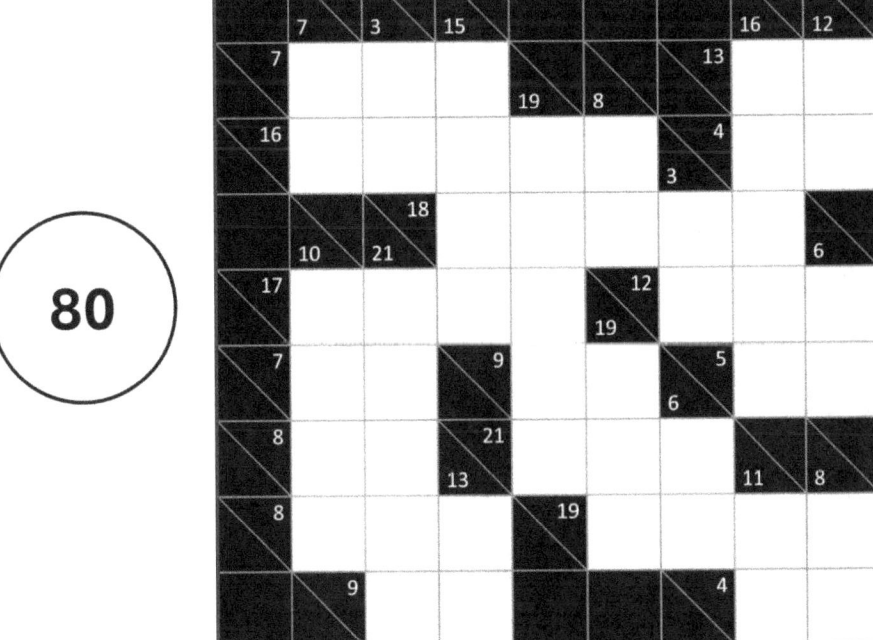

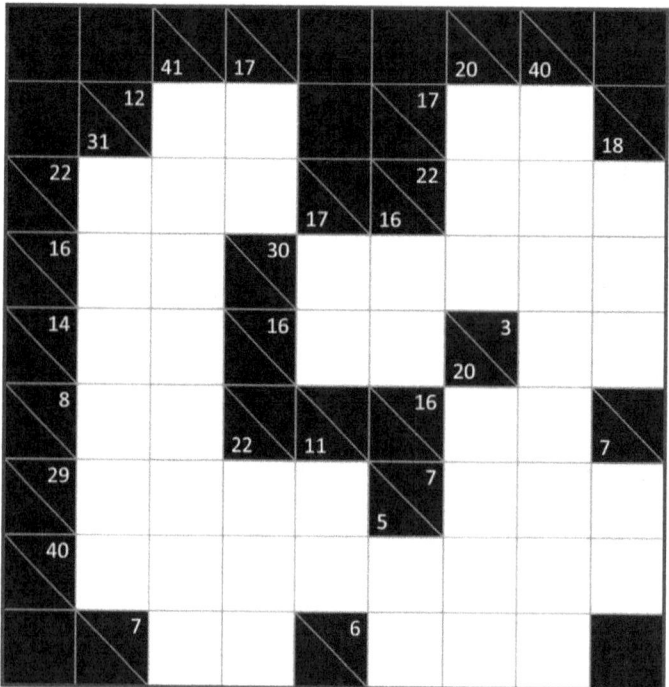

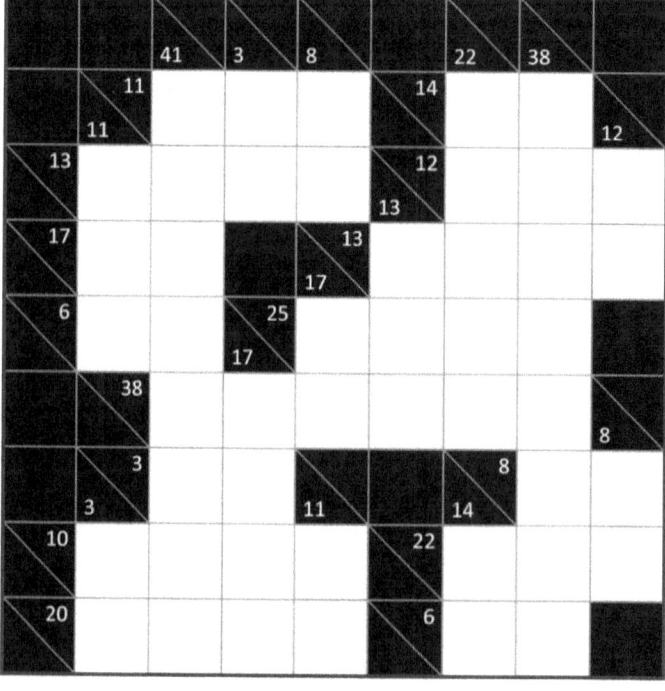

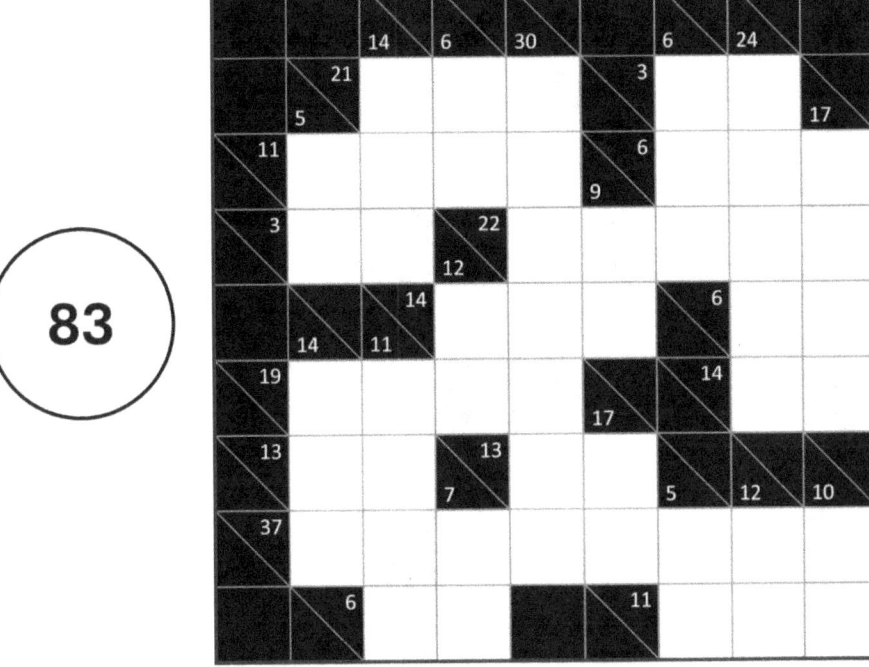

83

84

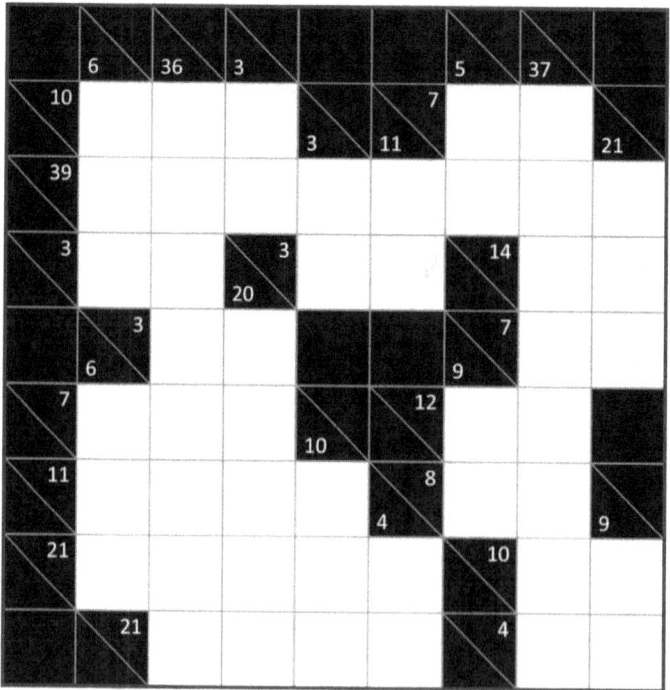

**87**

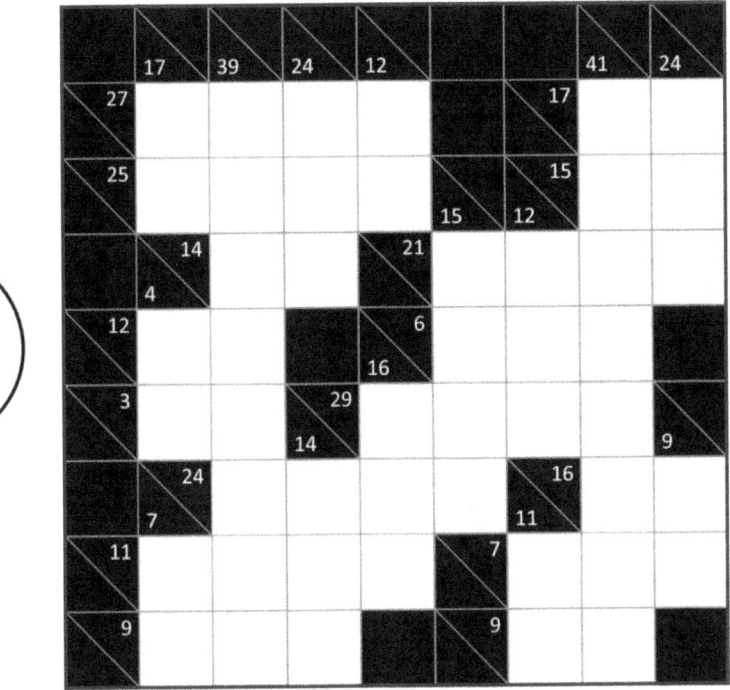

**88**

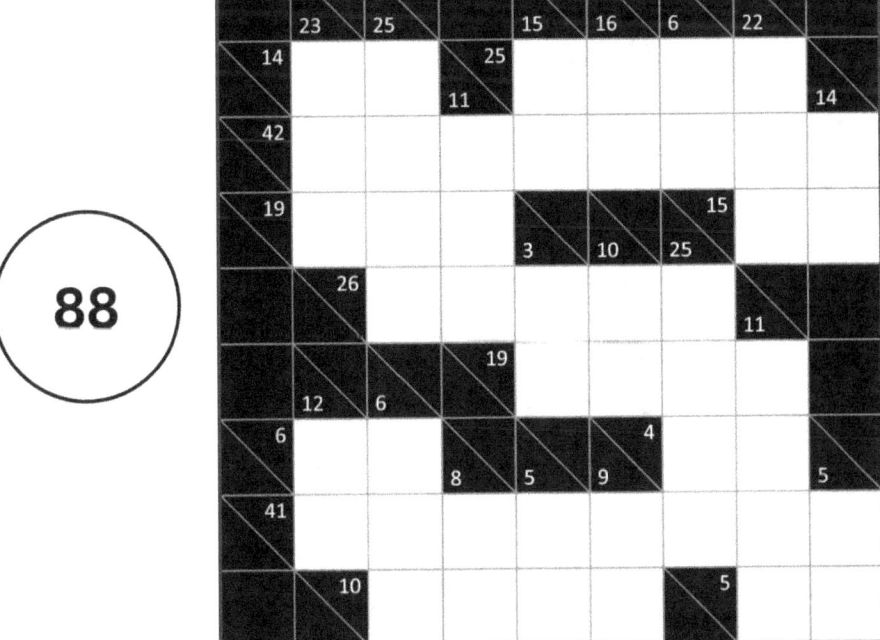

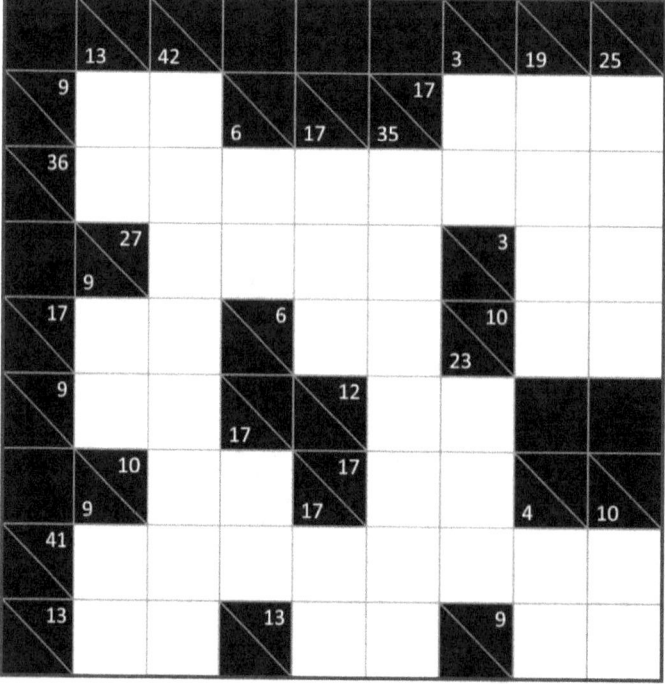

**91**

**92**

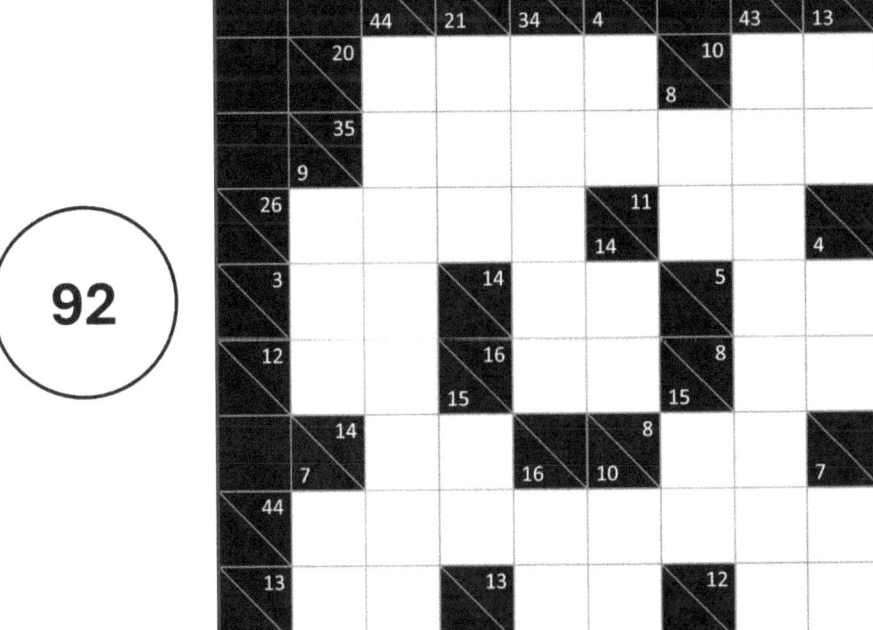

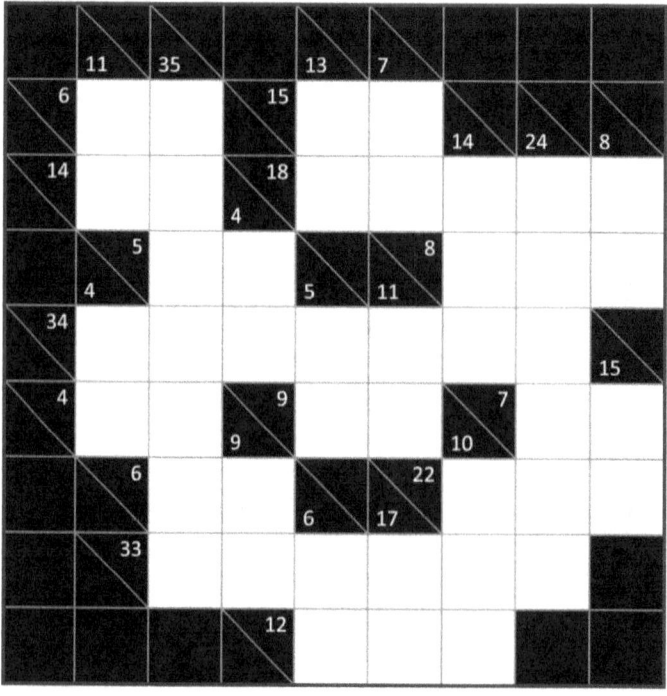

95

96

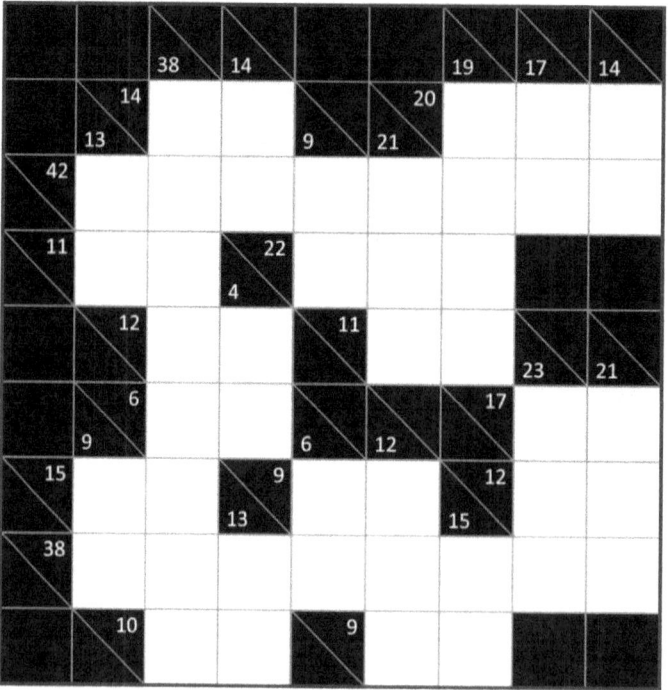

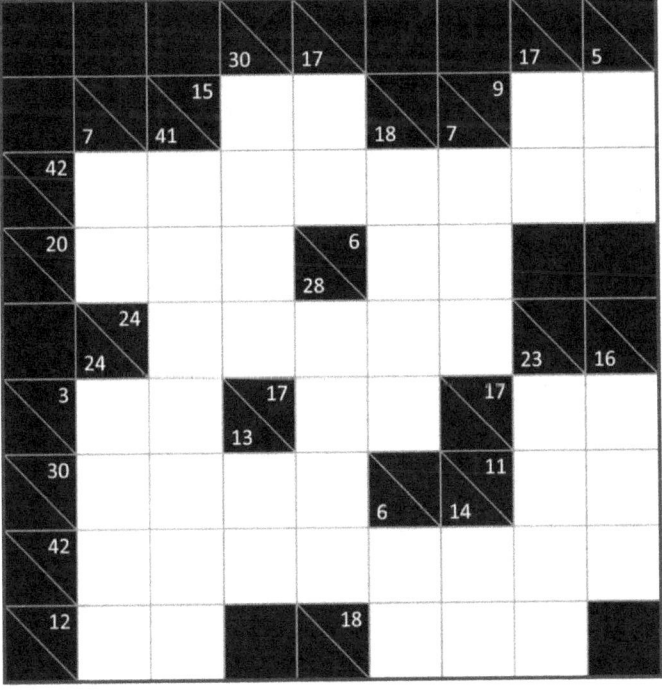

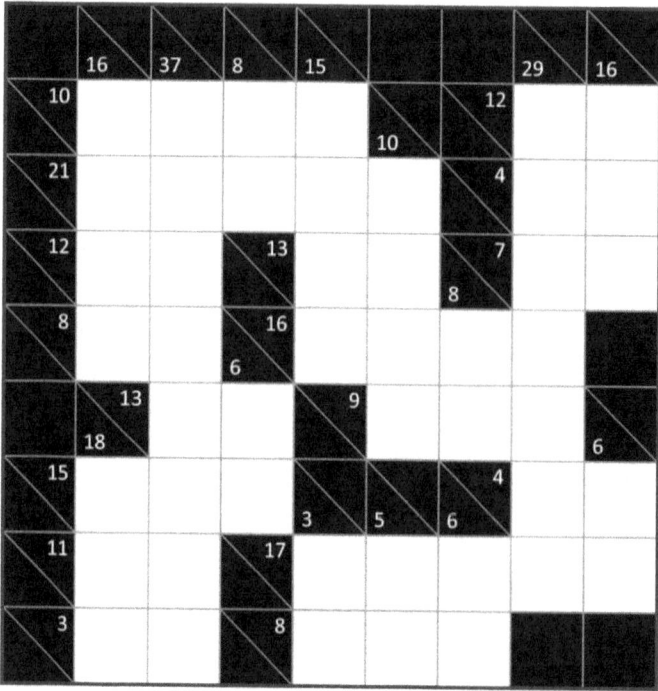

103

104

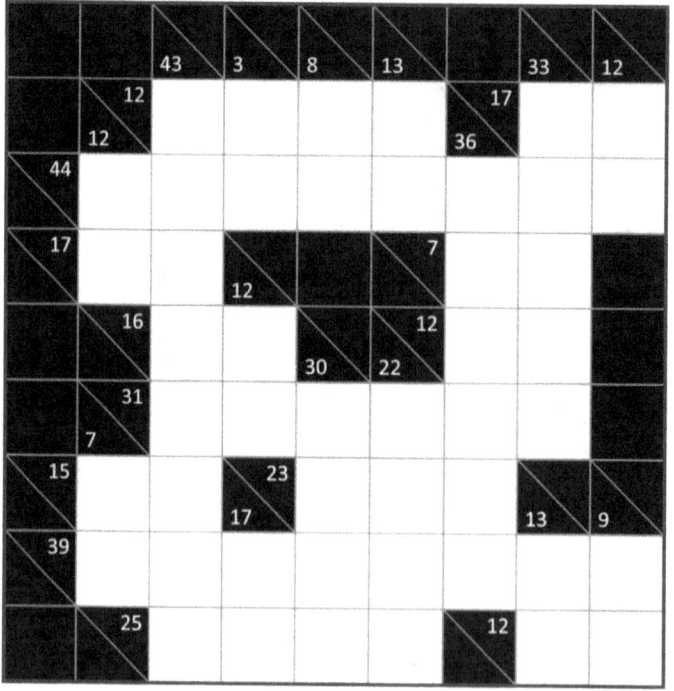

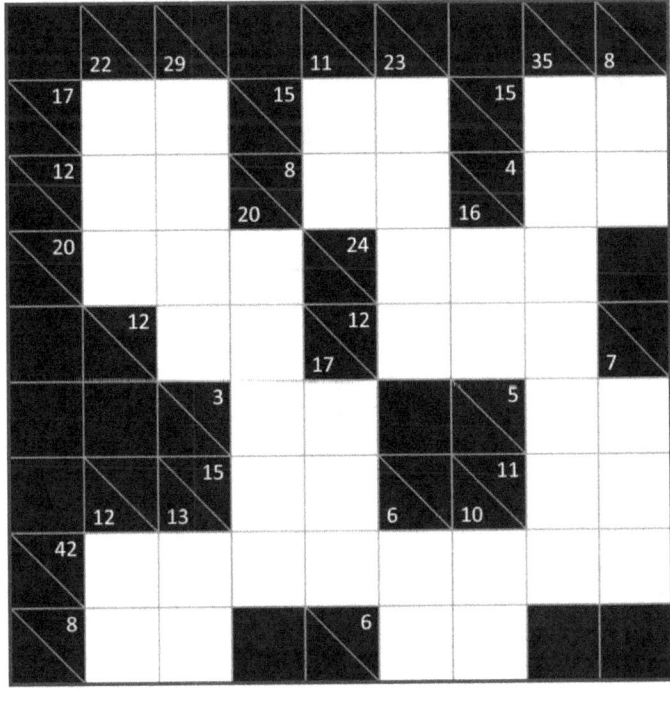

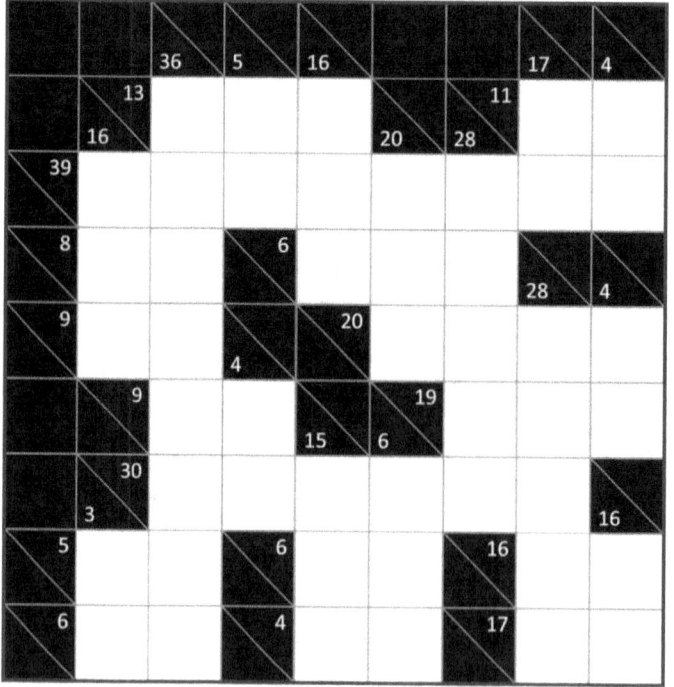

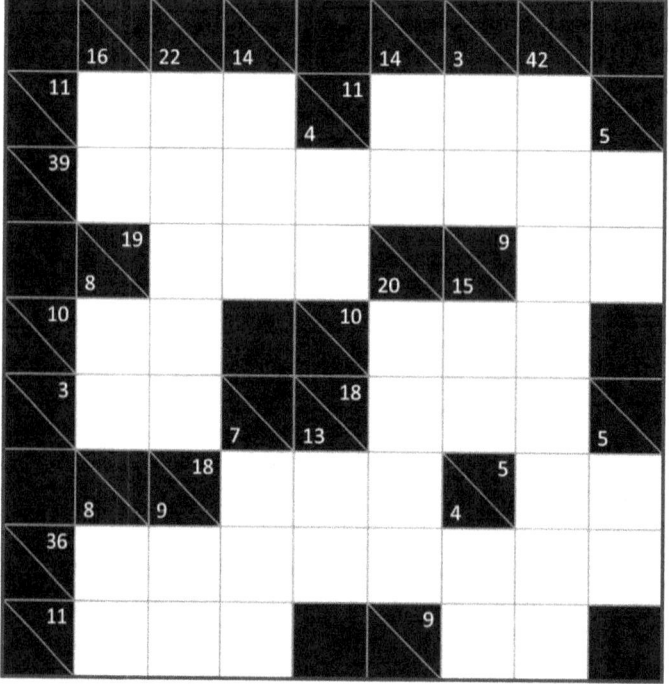

115

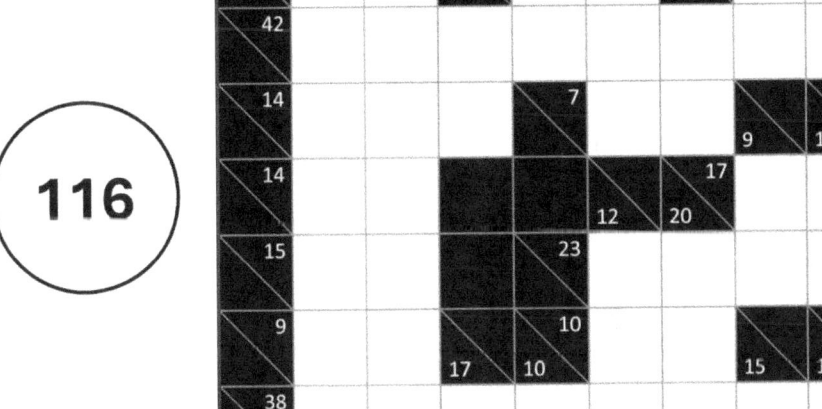

116

**119**

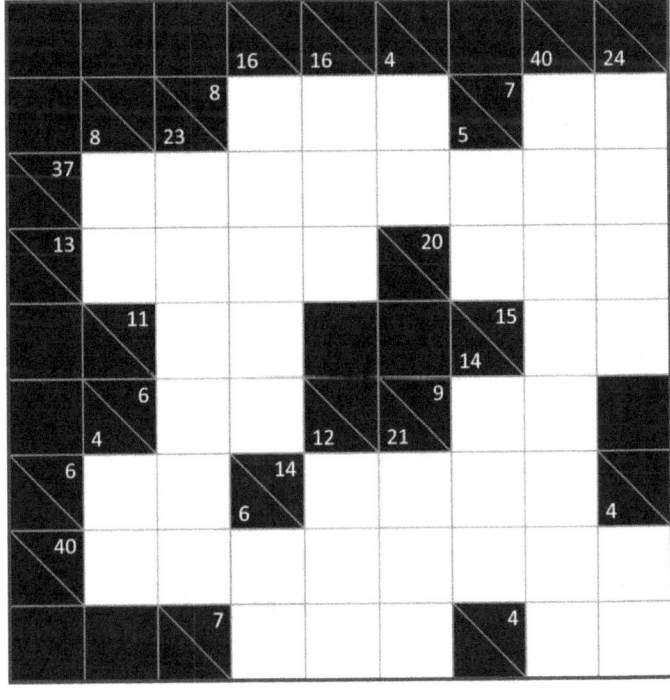

**120**

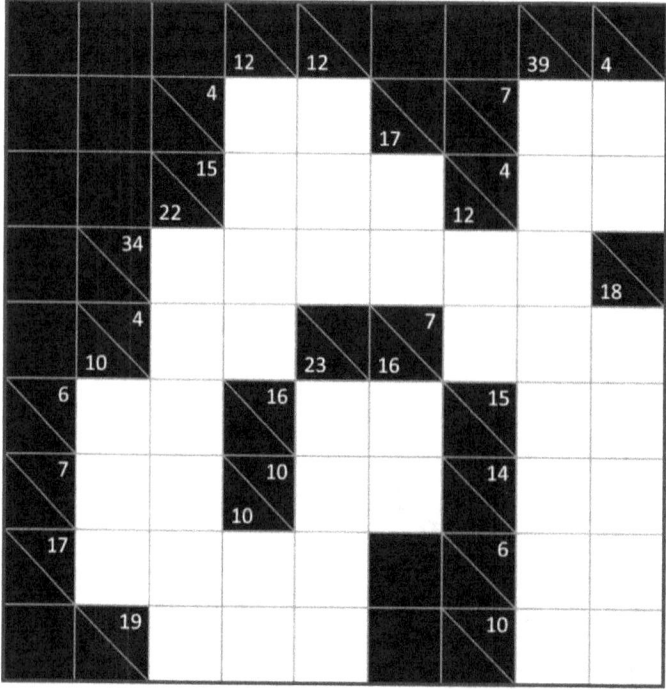

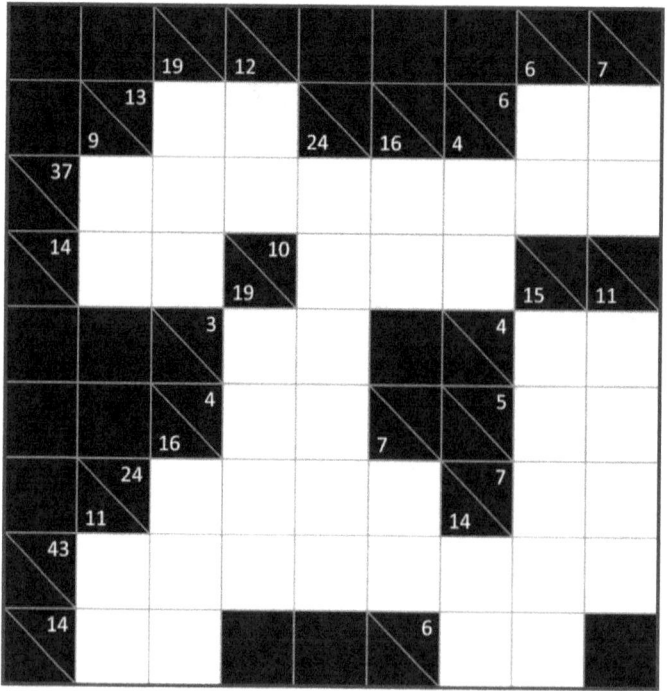

127

128

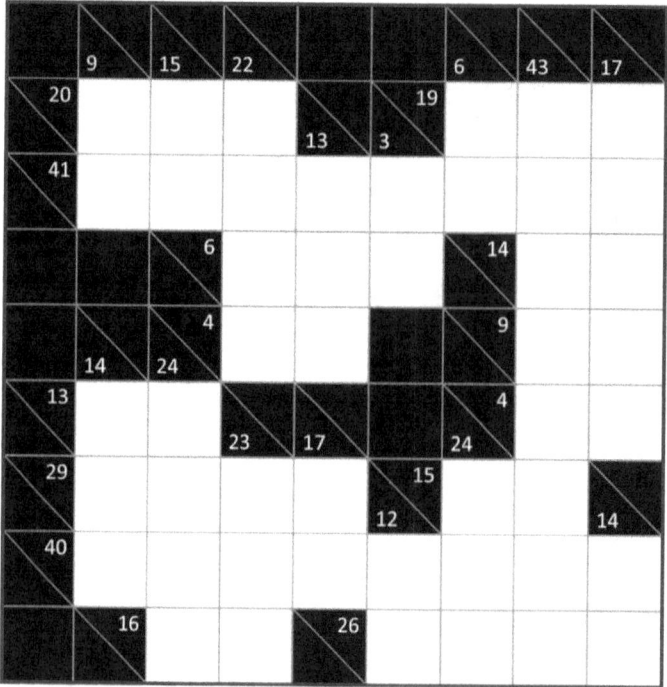

**131**

**132**

135

136

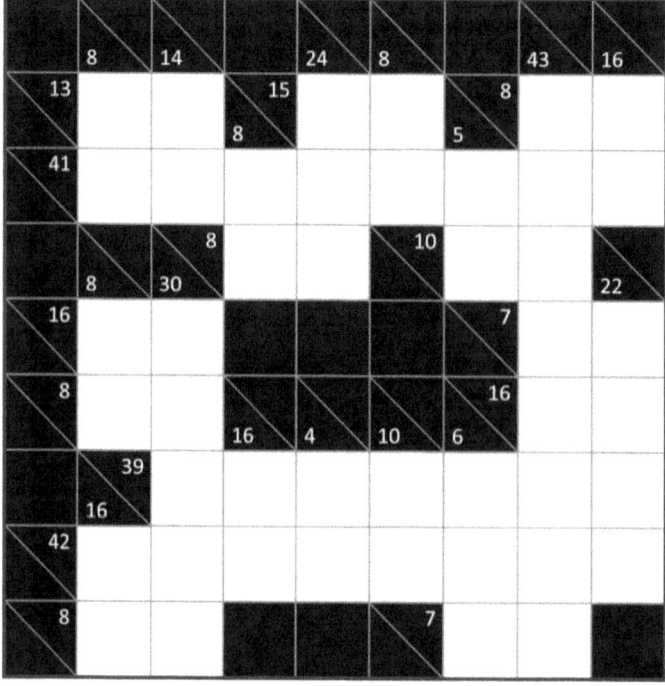

**139**

**140**

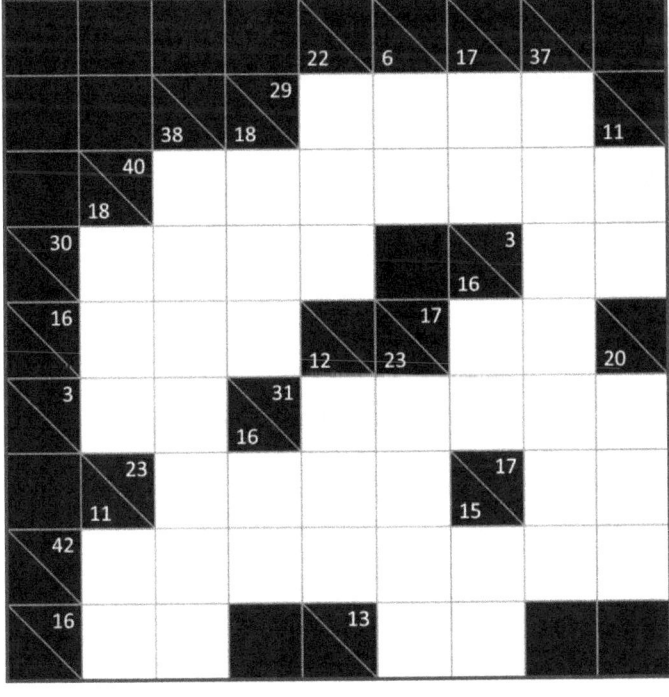

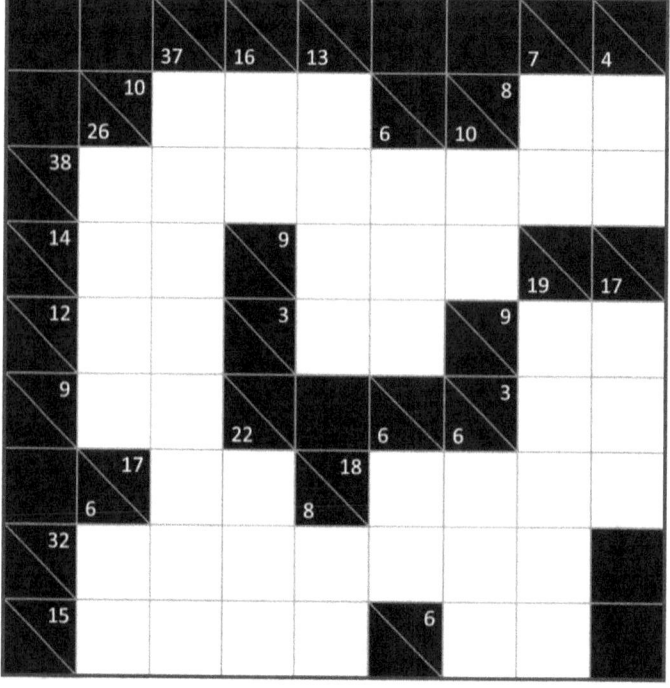

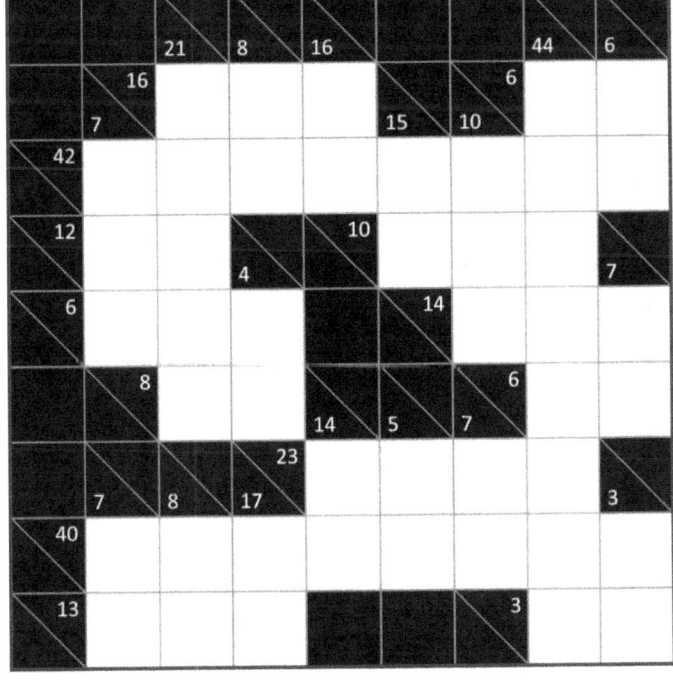

**143**

**144**

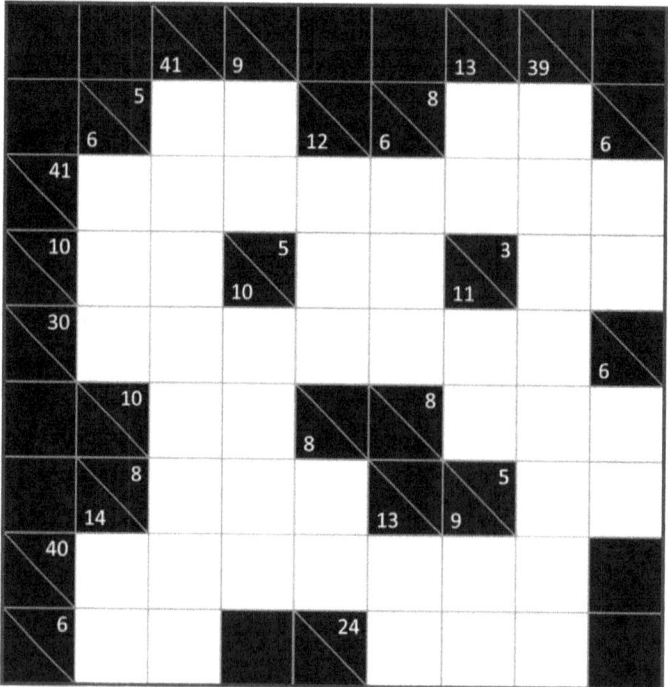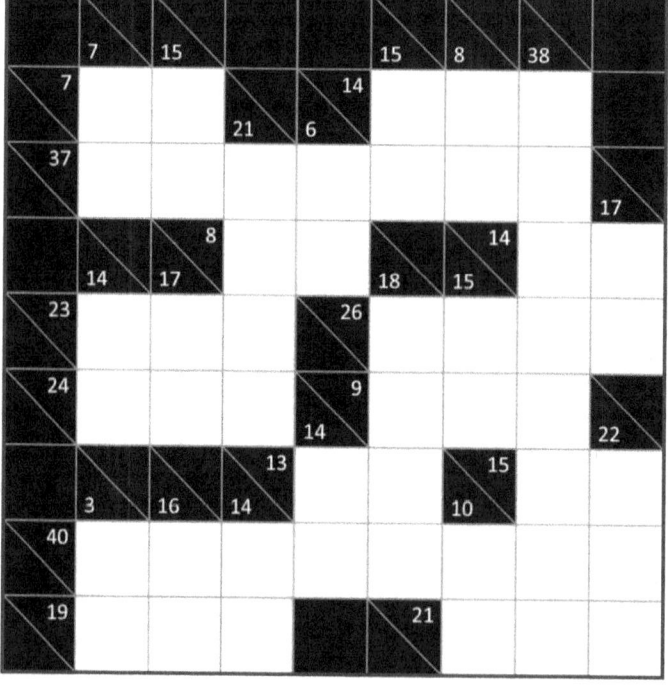

**147**

**148**

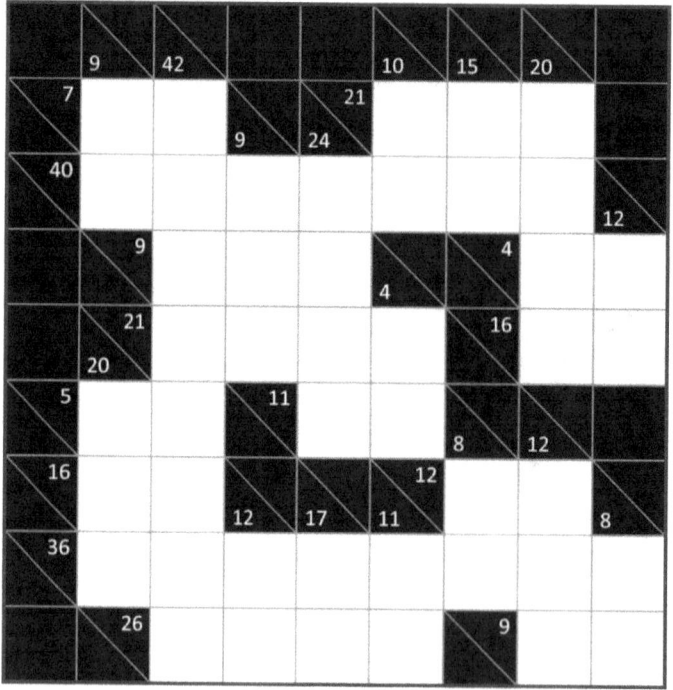

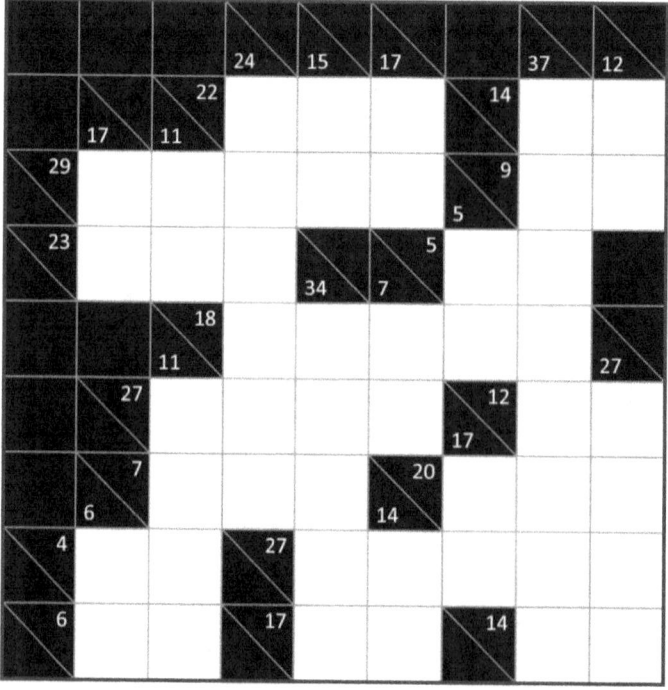

## 151

## 152

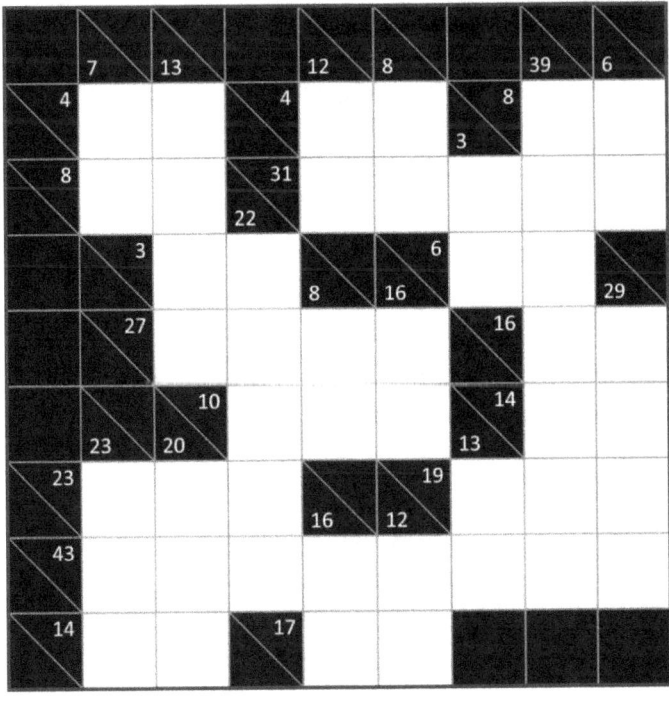

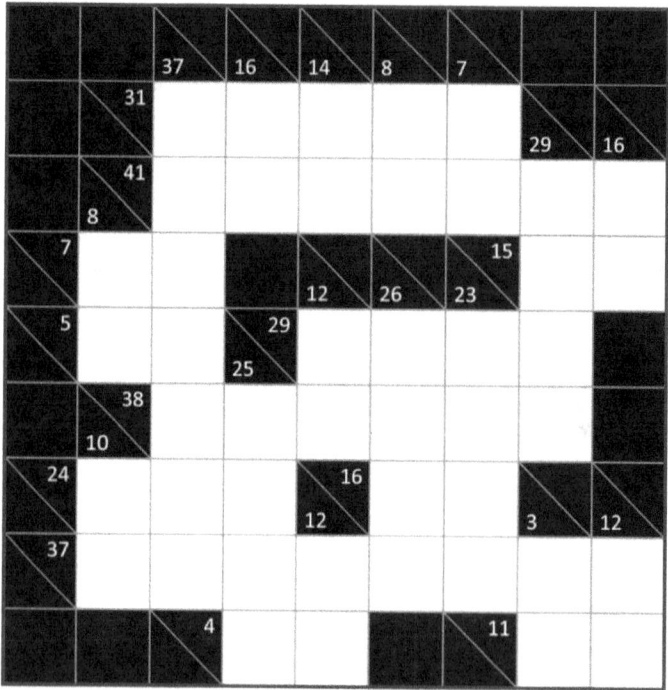

153

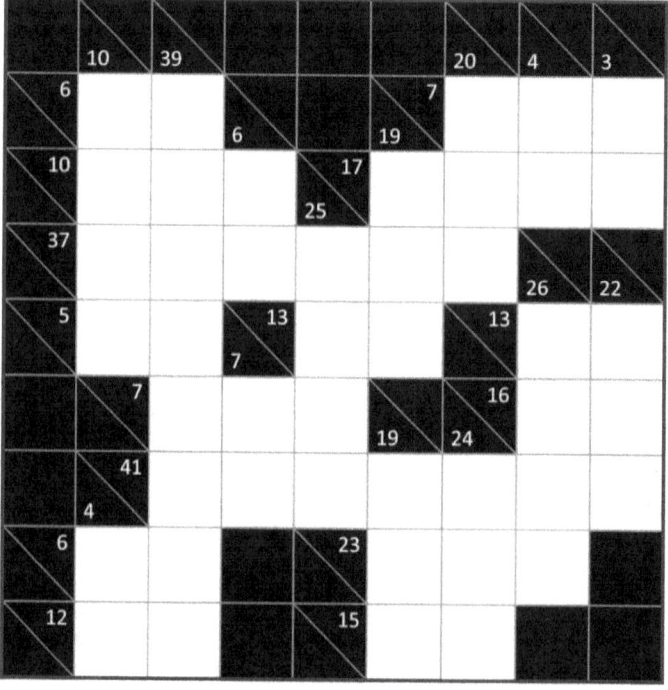

154

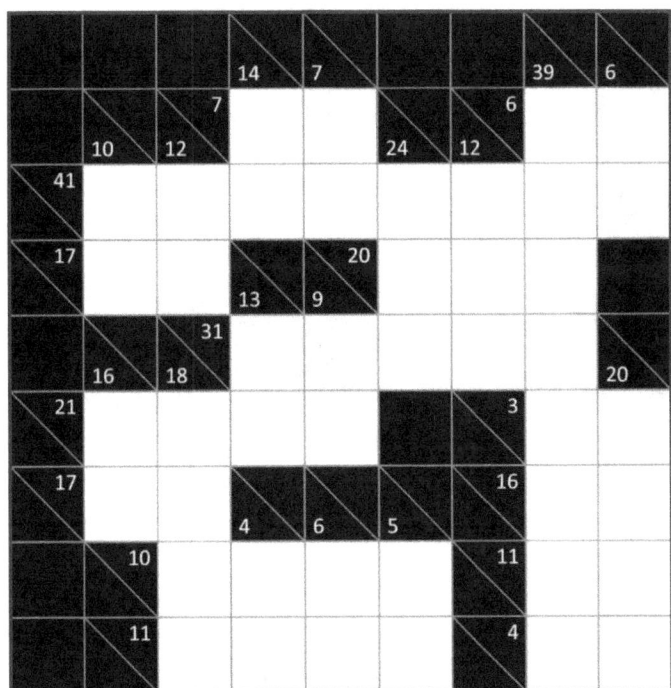

**155**

**156**

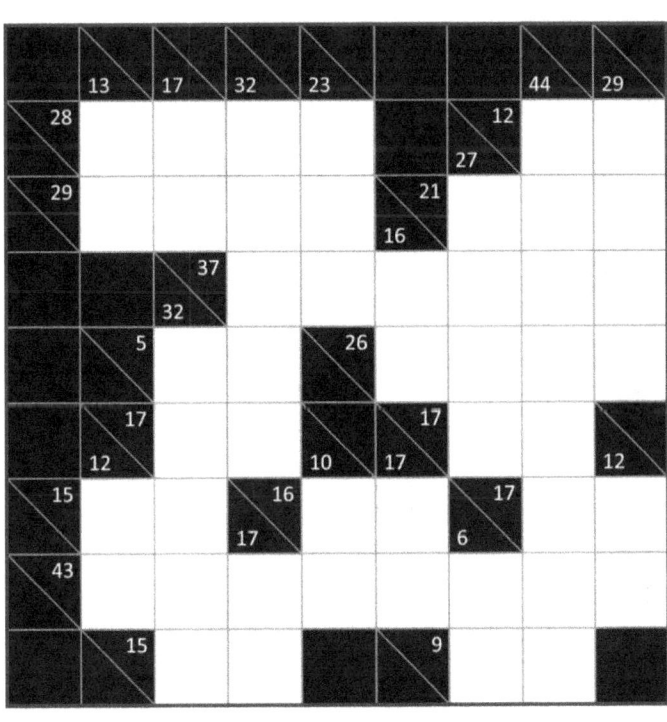

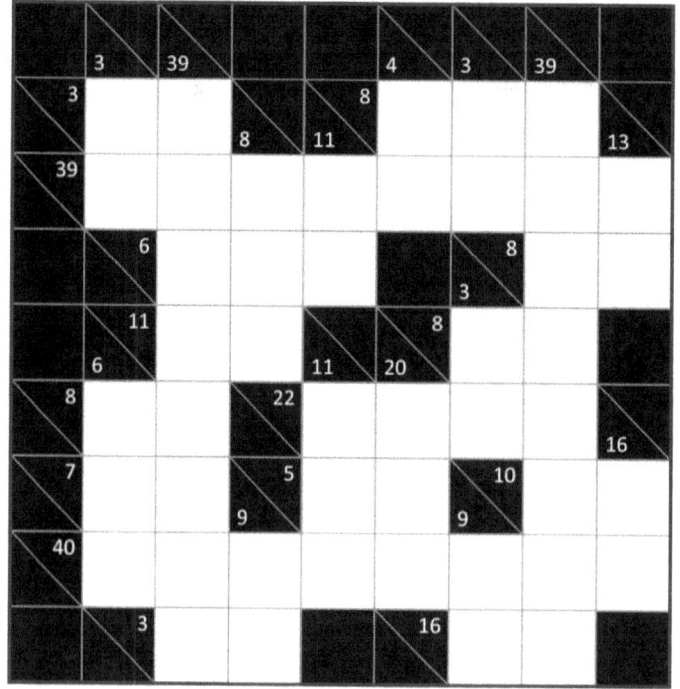

**159**

**160**

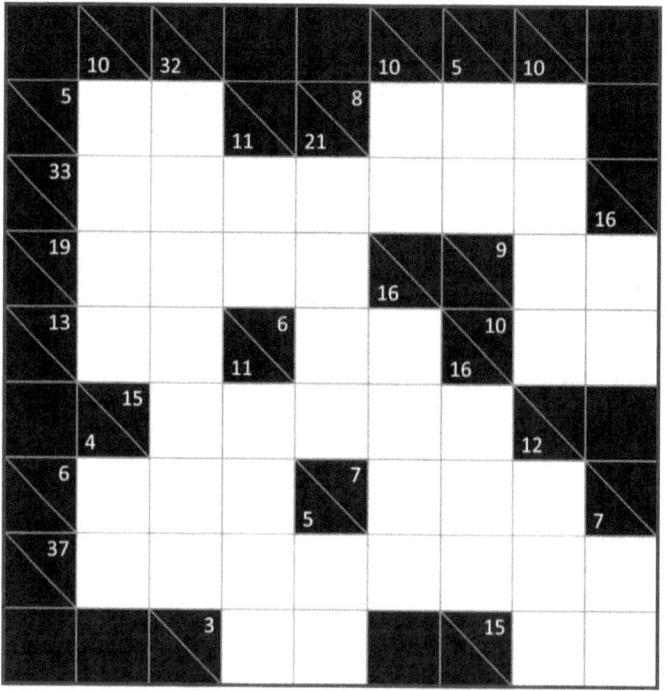

**163**

**164**

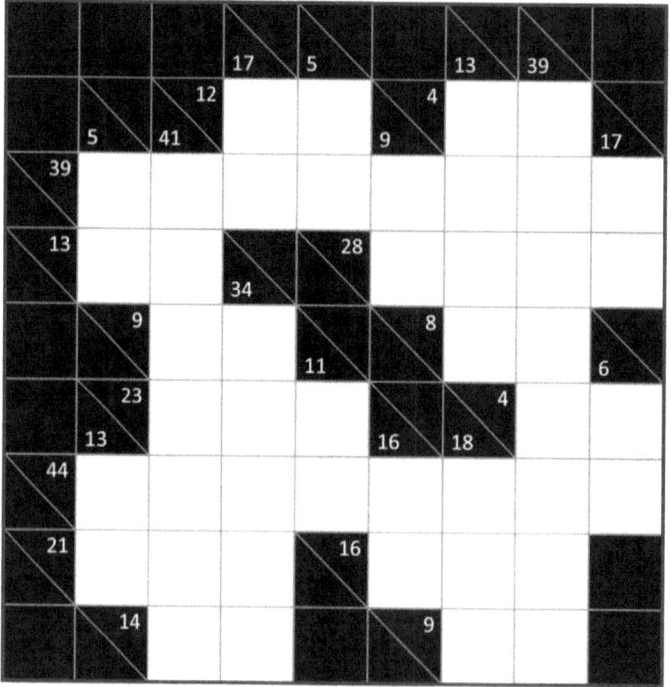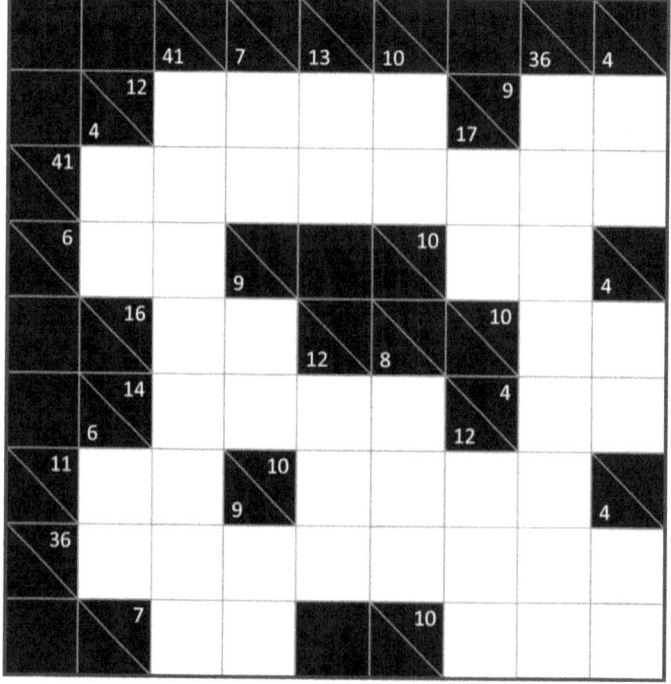

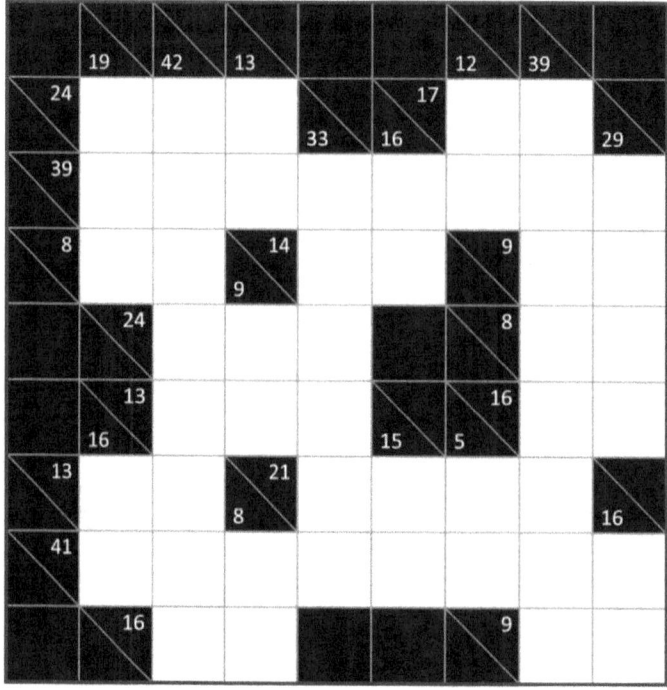

171

172

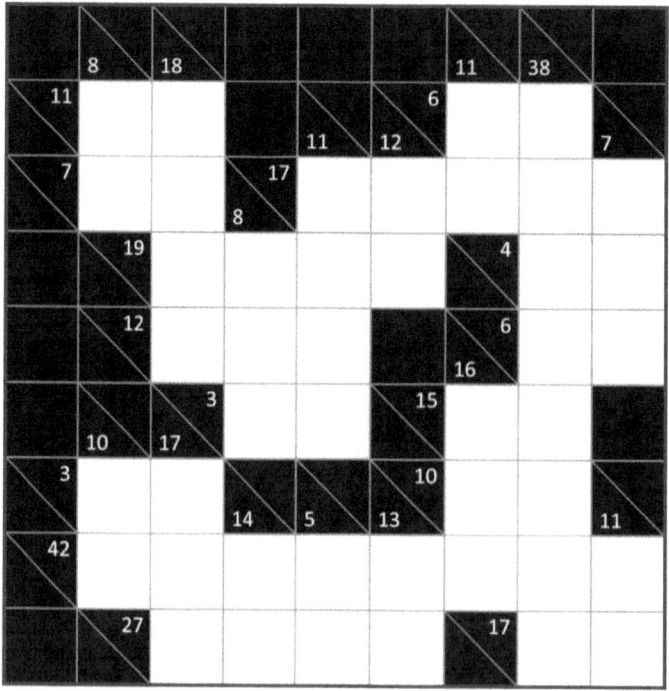

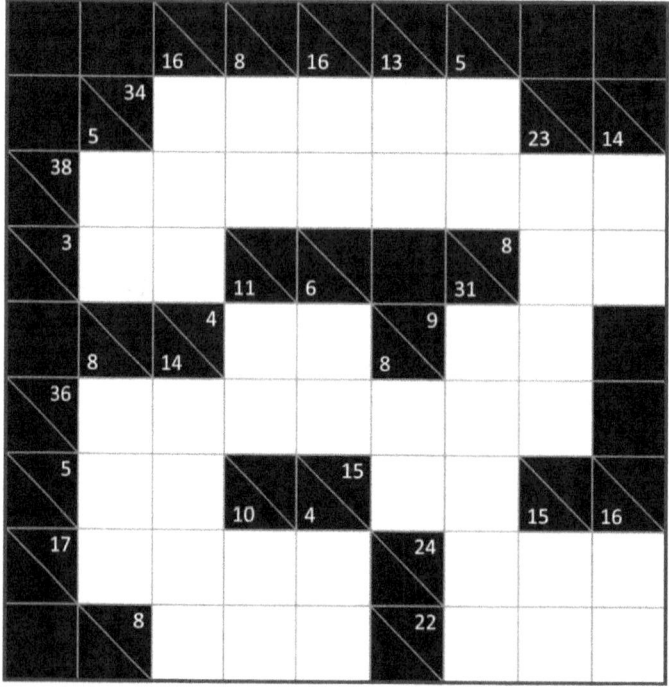

## 175

## 176

**179**

**180**

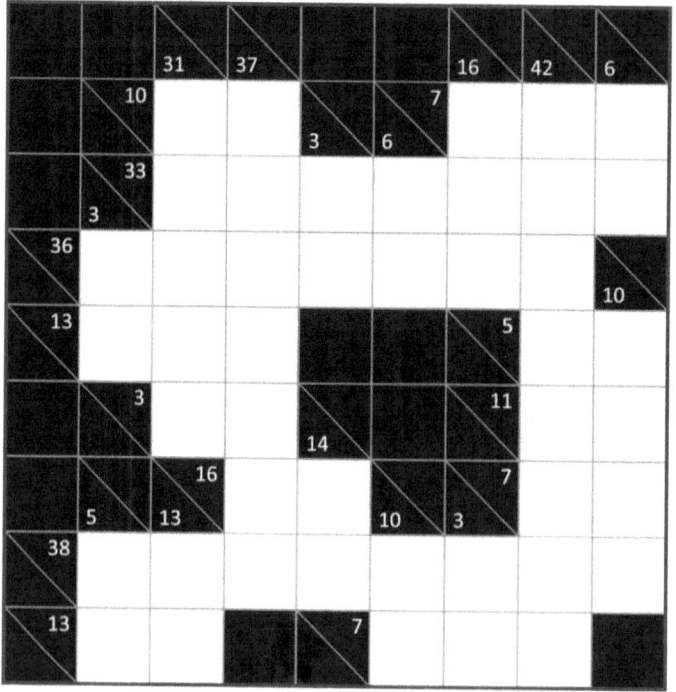

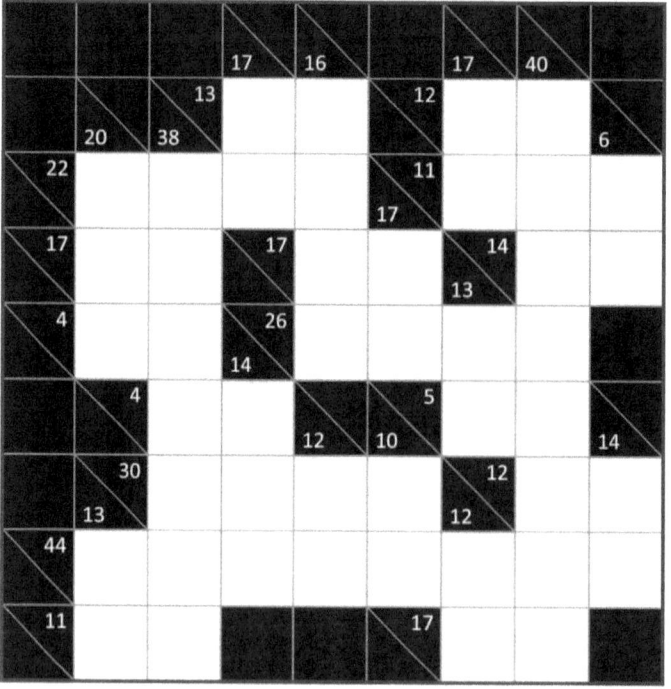

**187**

**188**

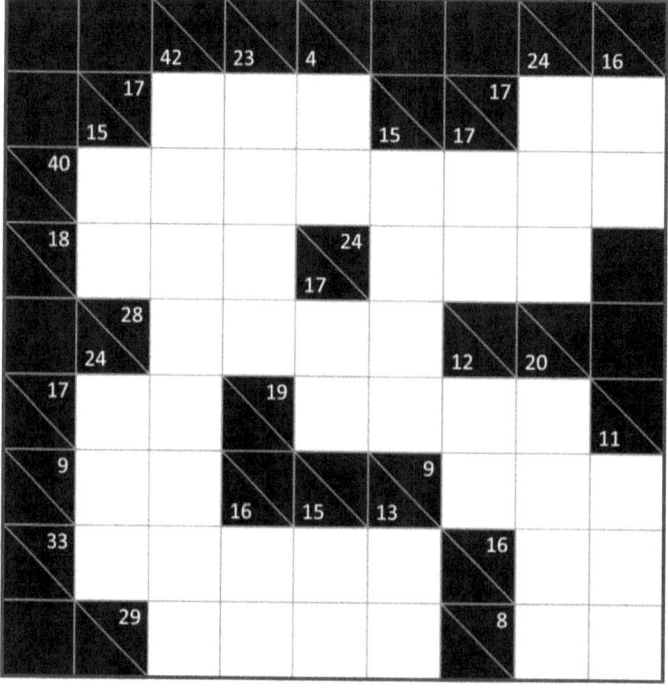

191

192

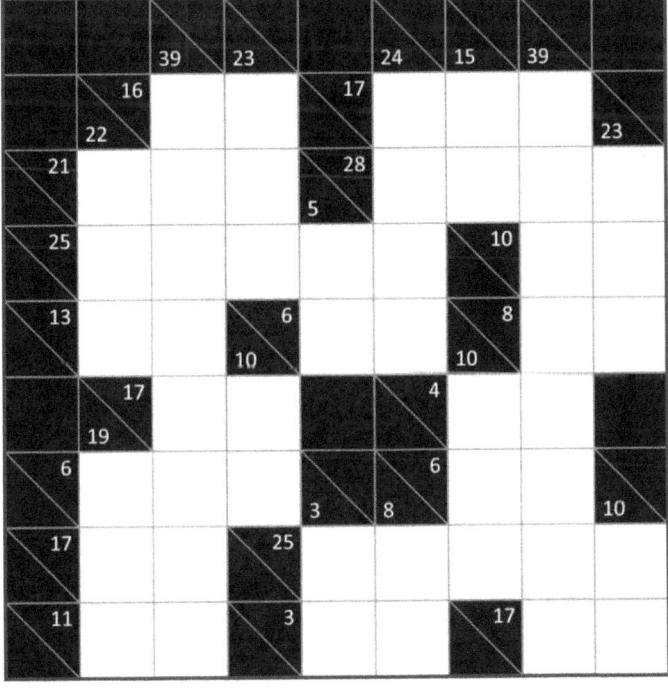

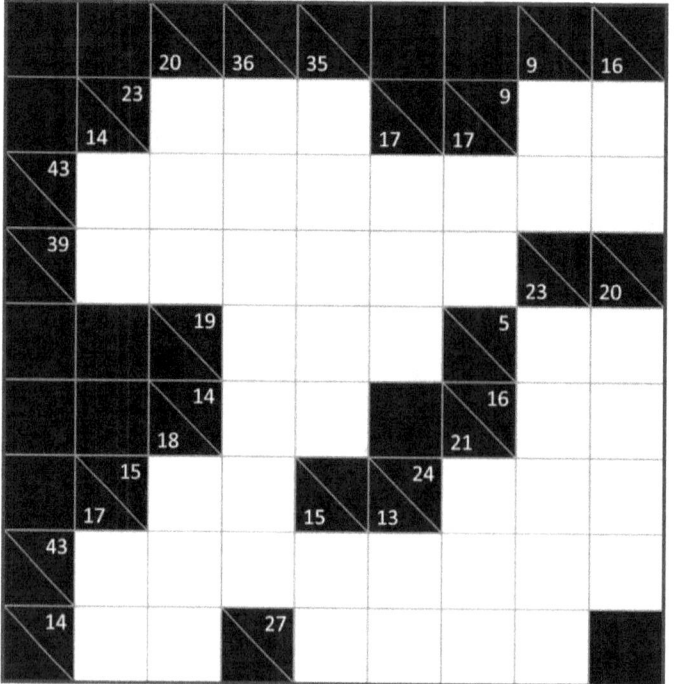

**195**

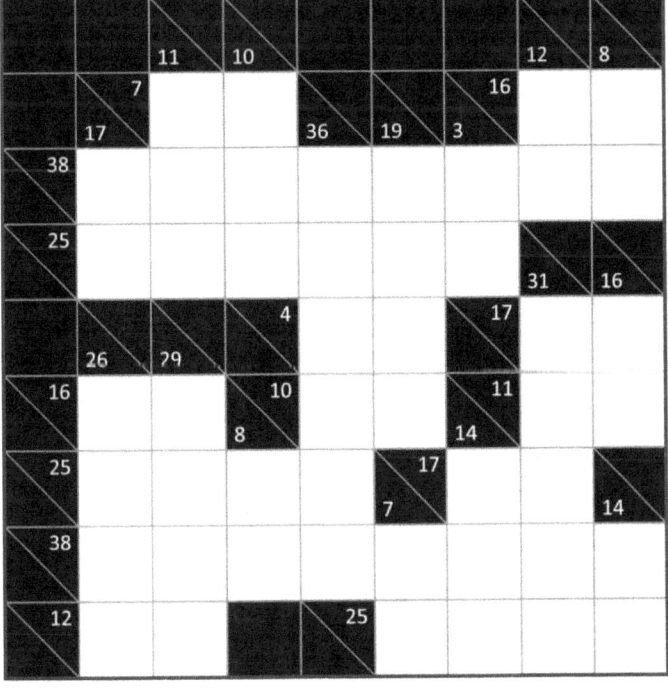

**196**

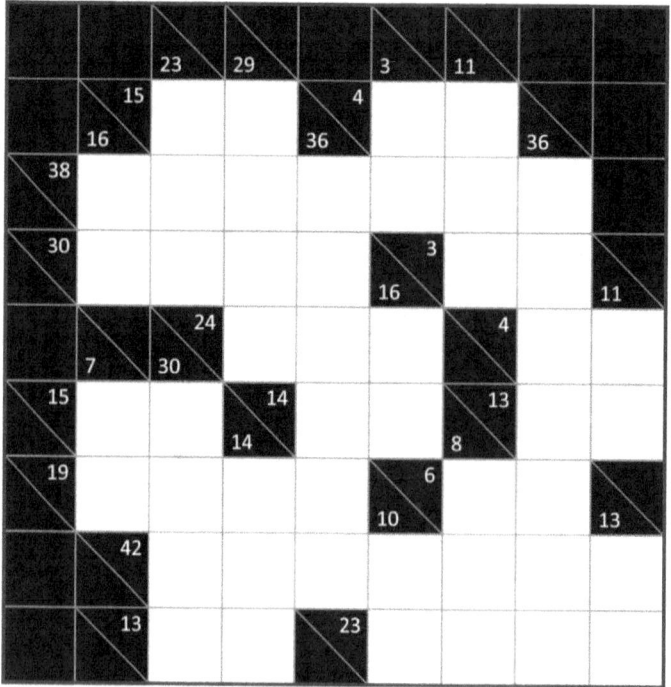

**199**

**200**

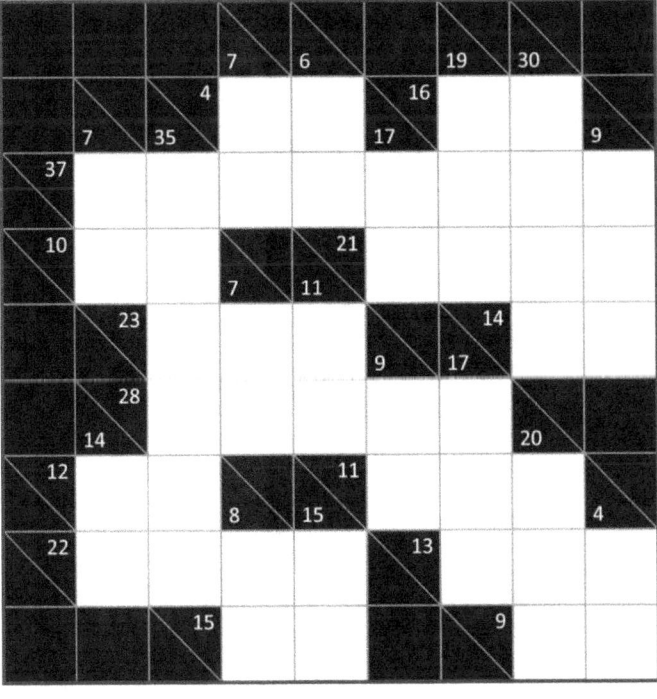

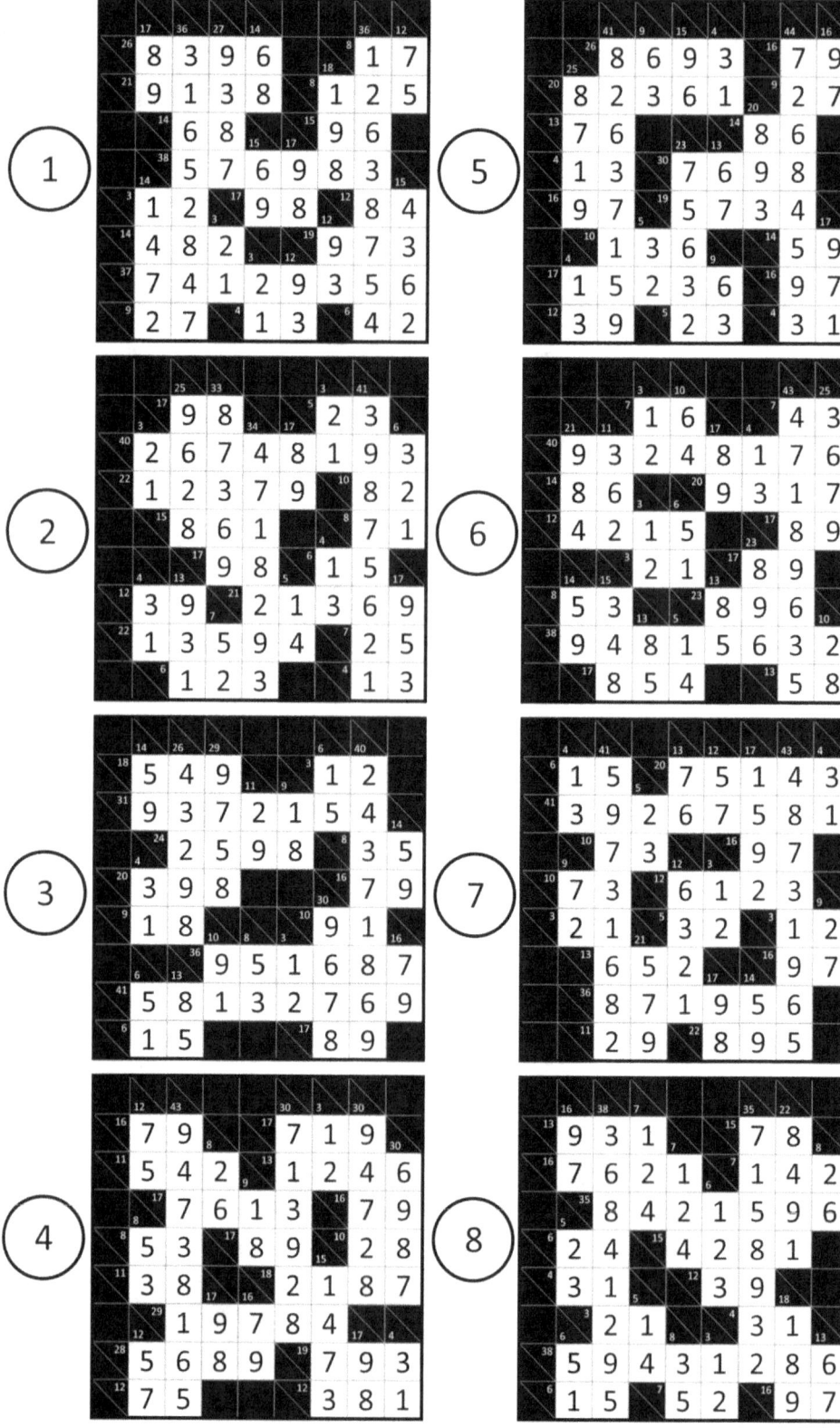

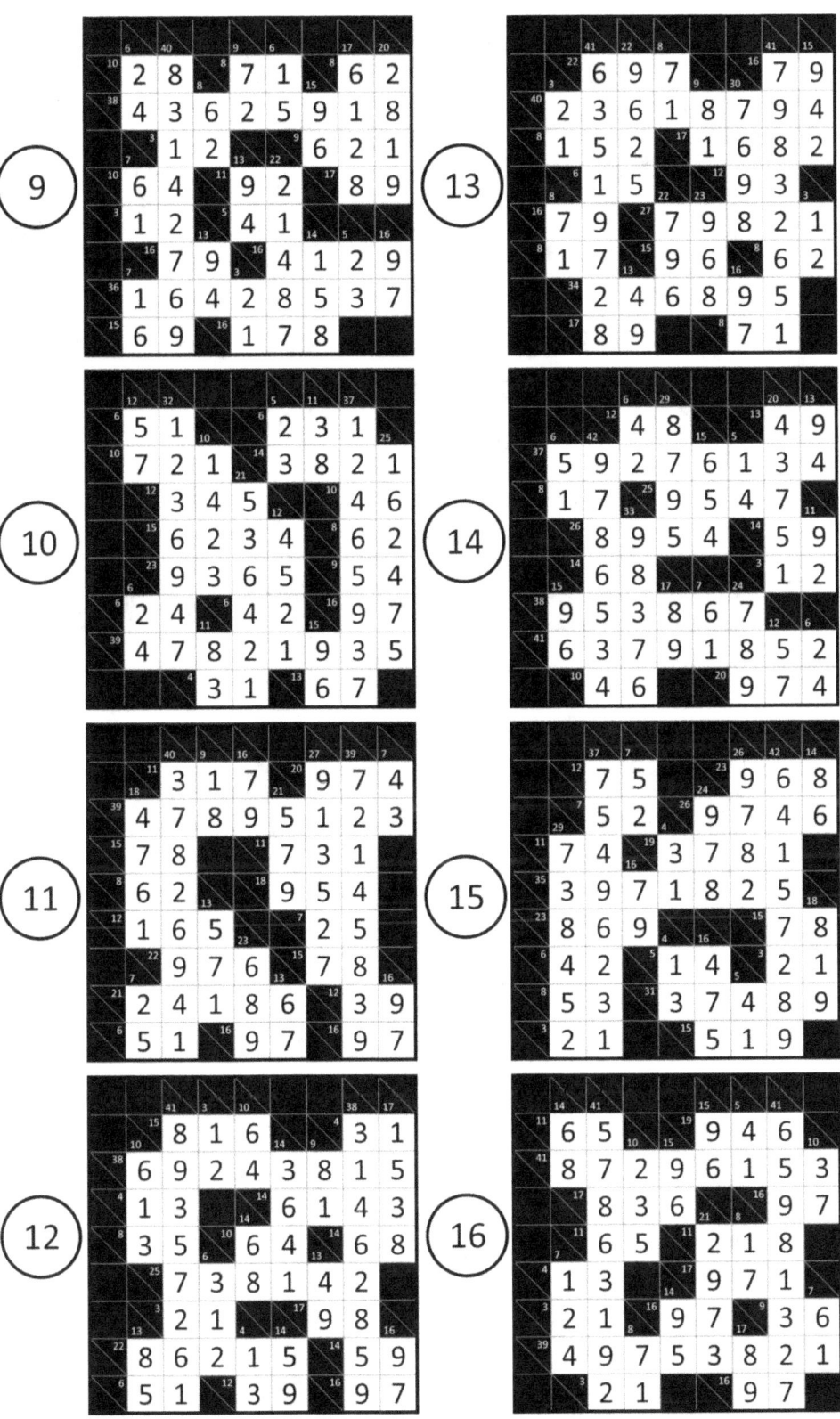

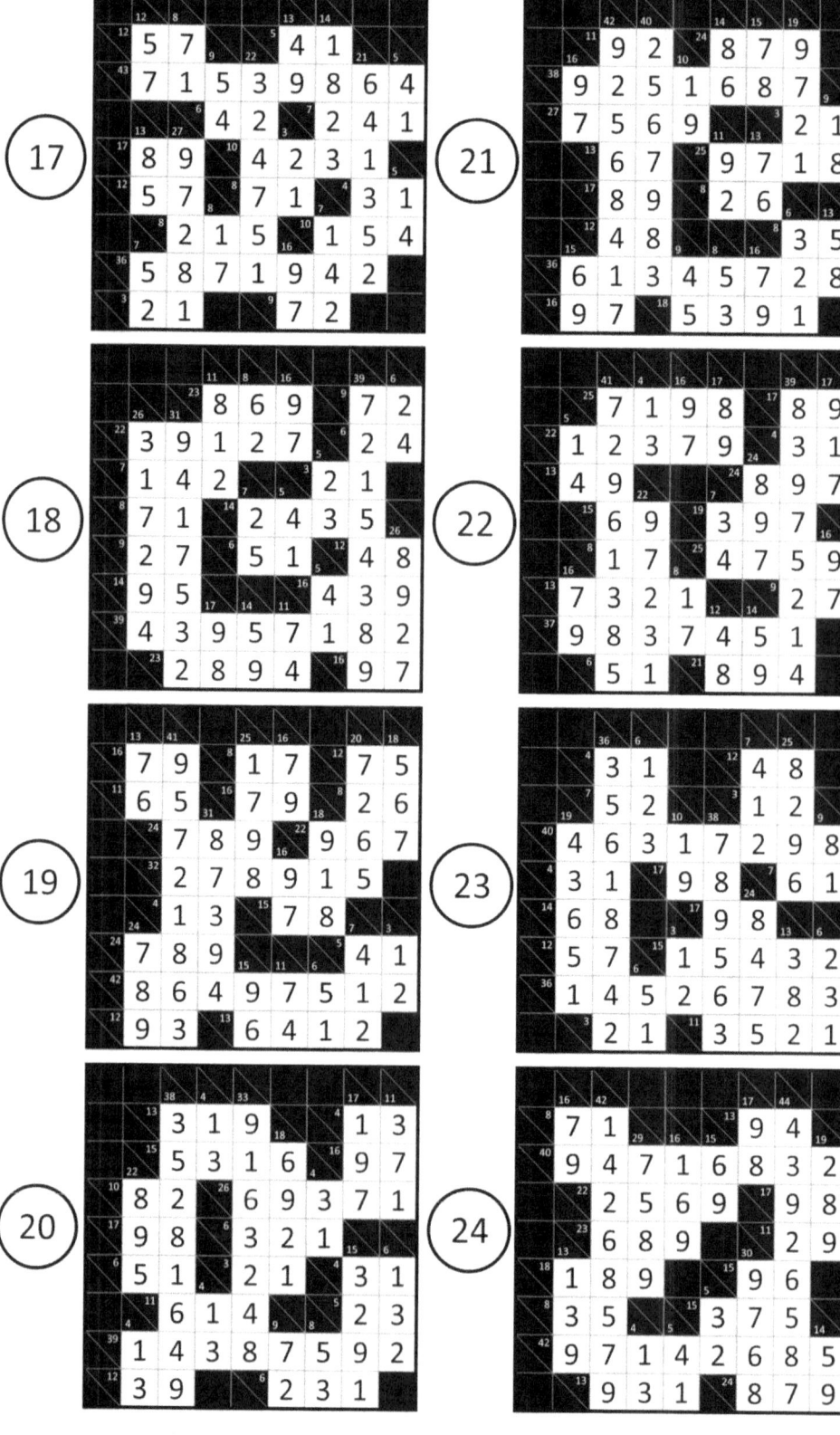

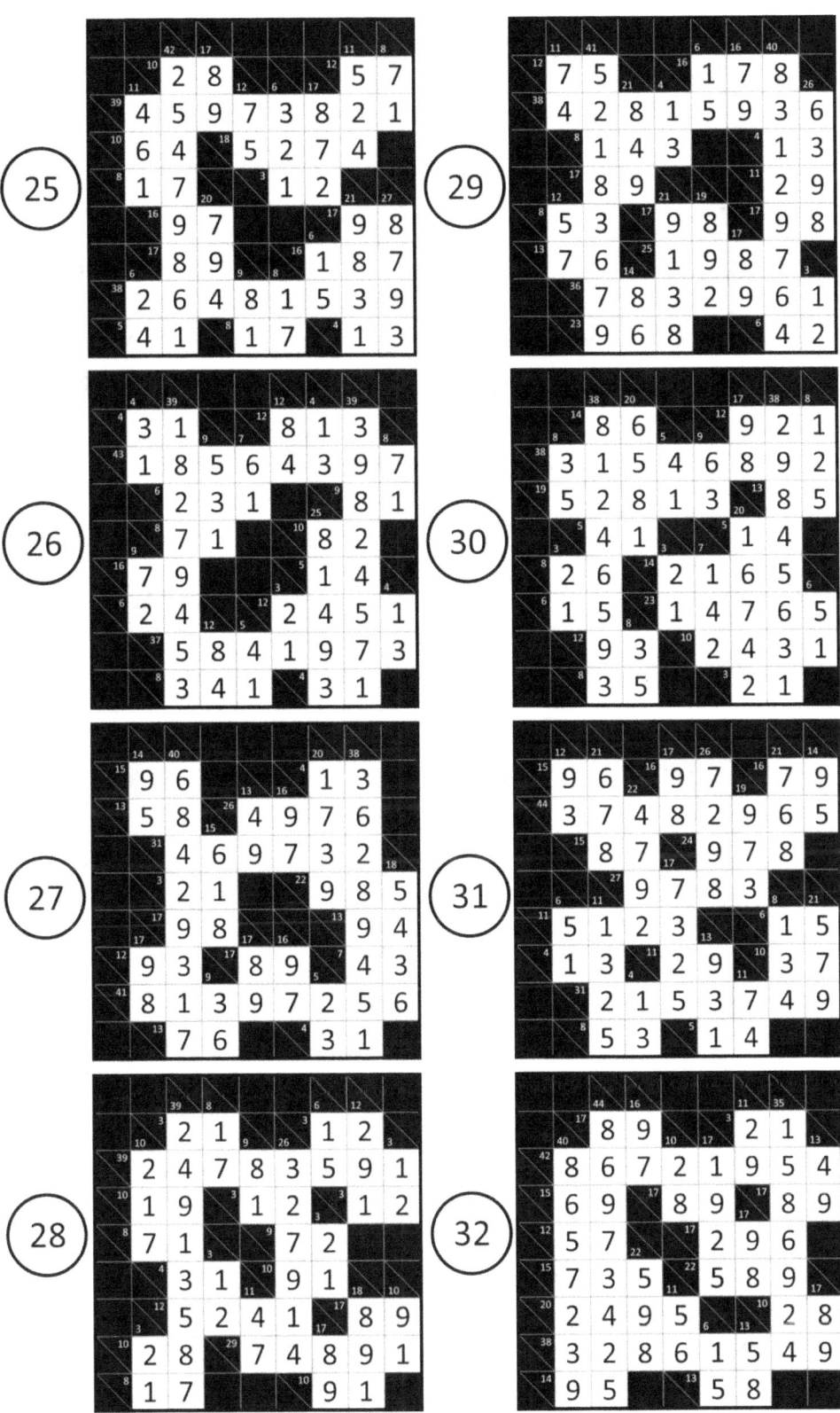

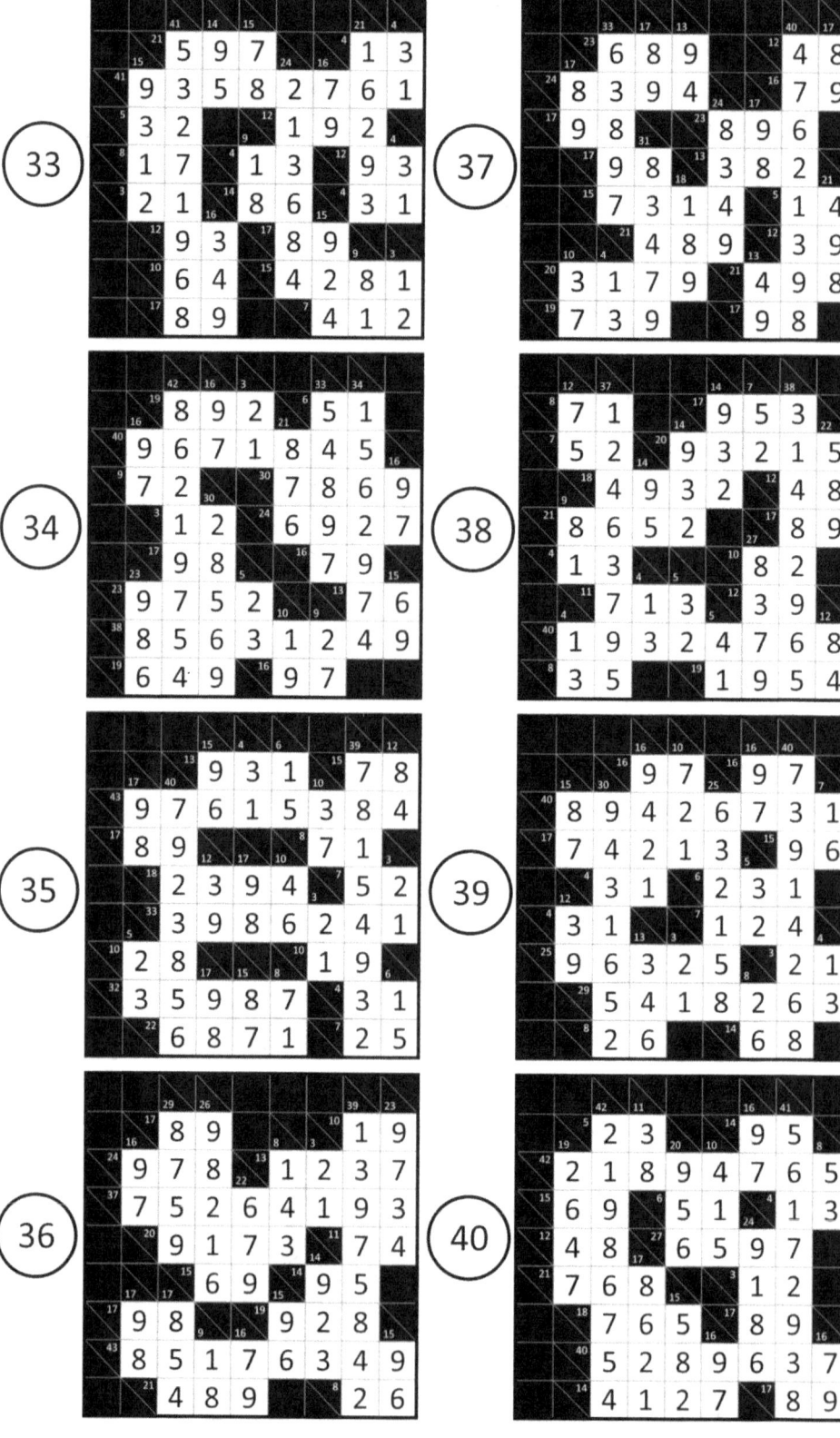

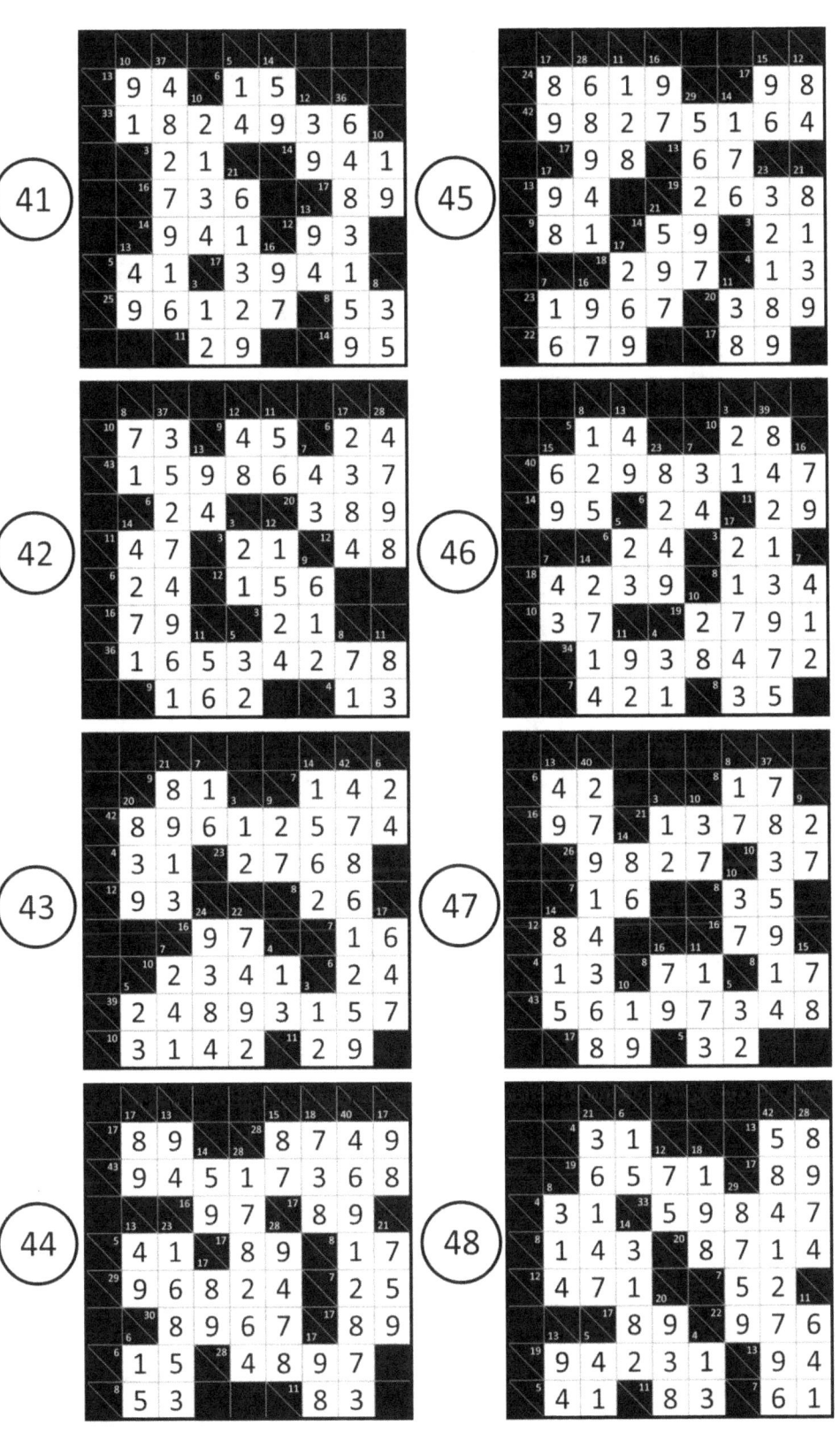

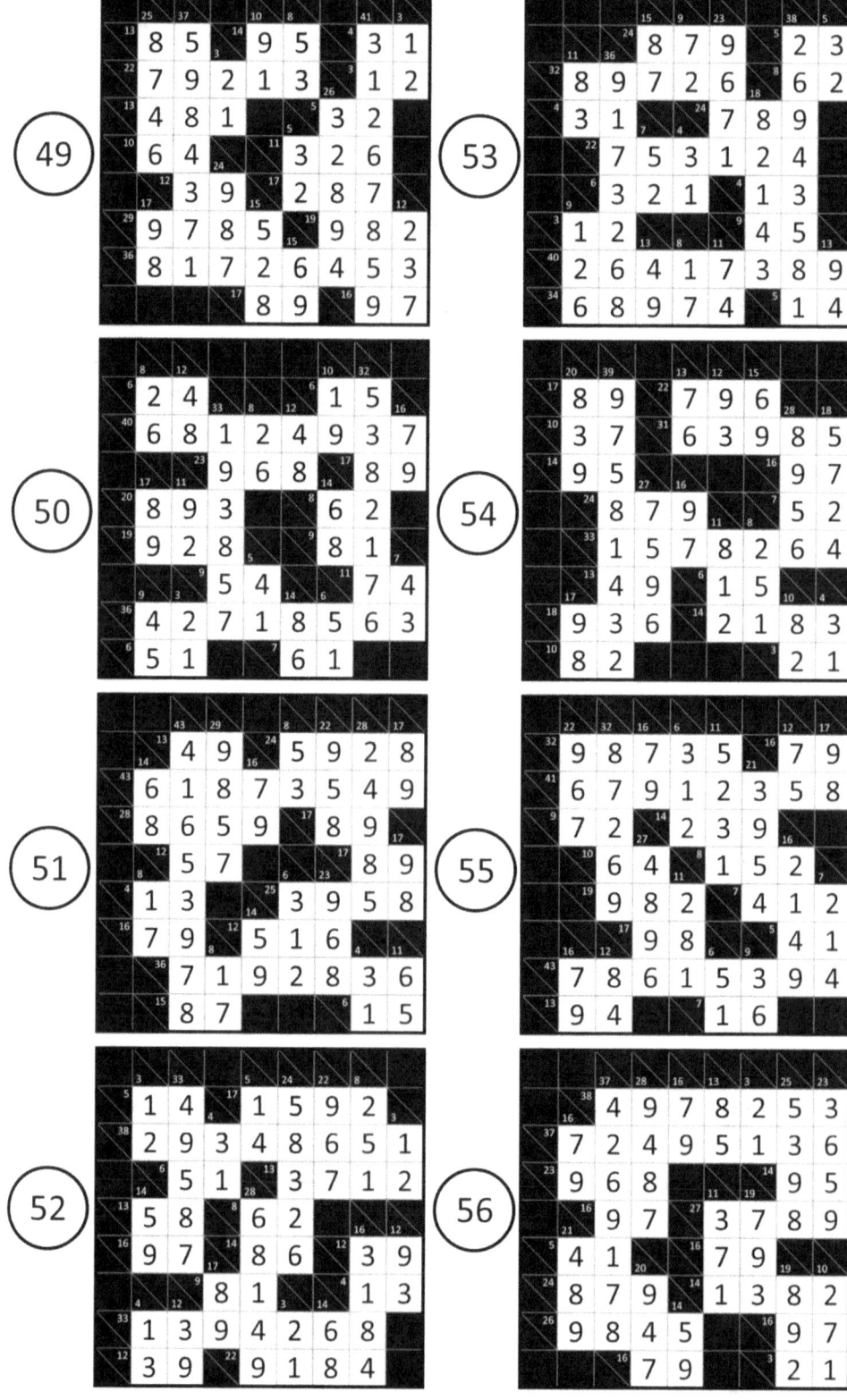

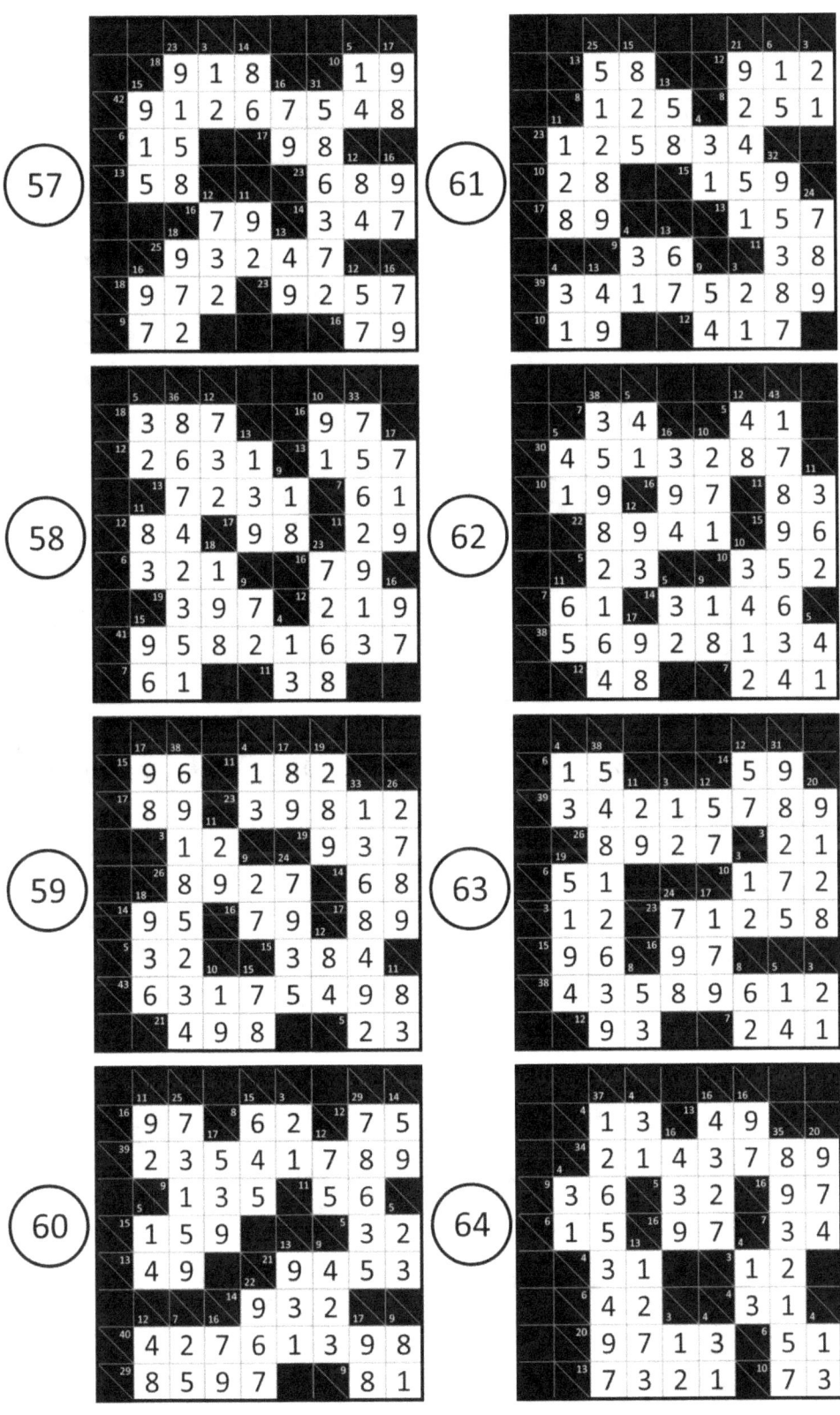

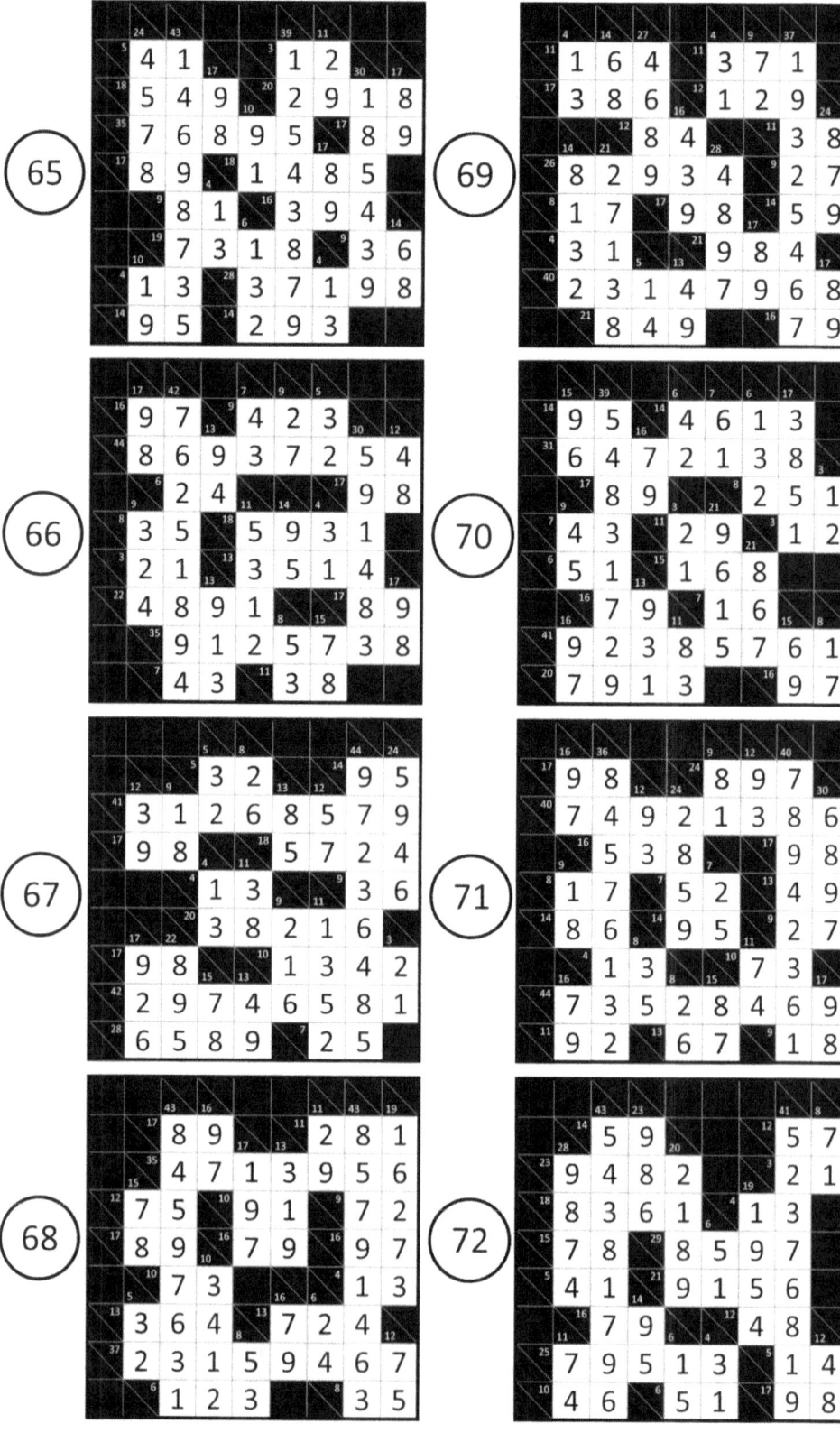

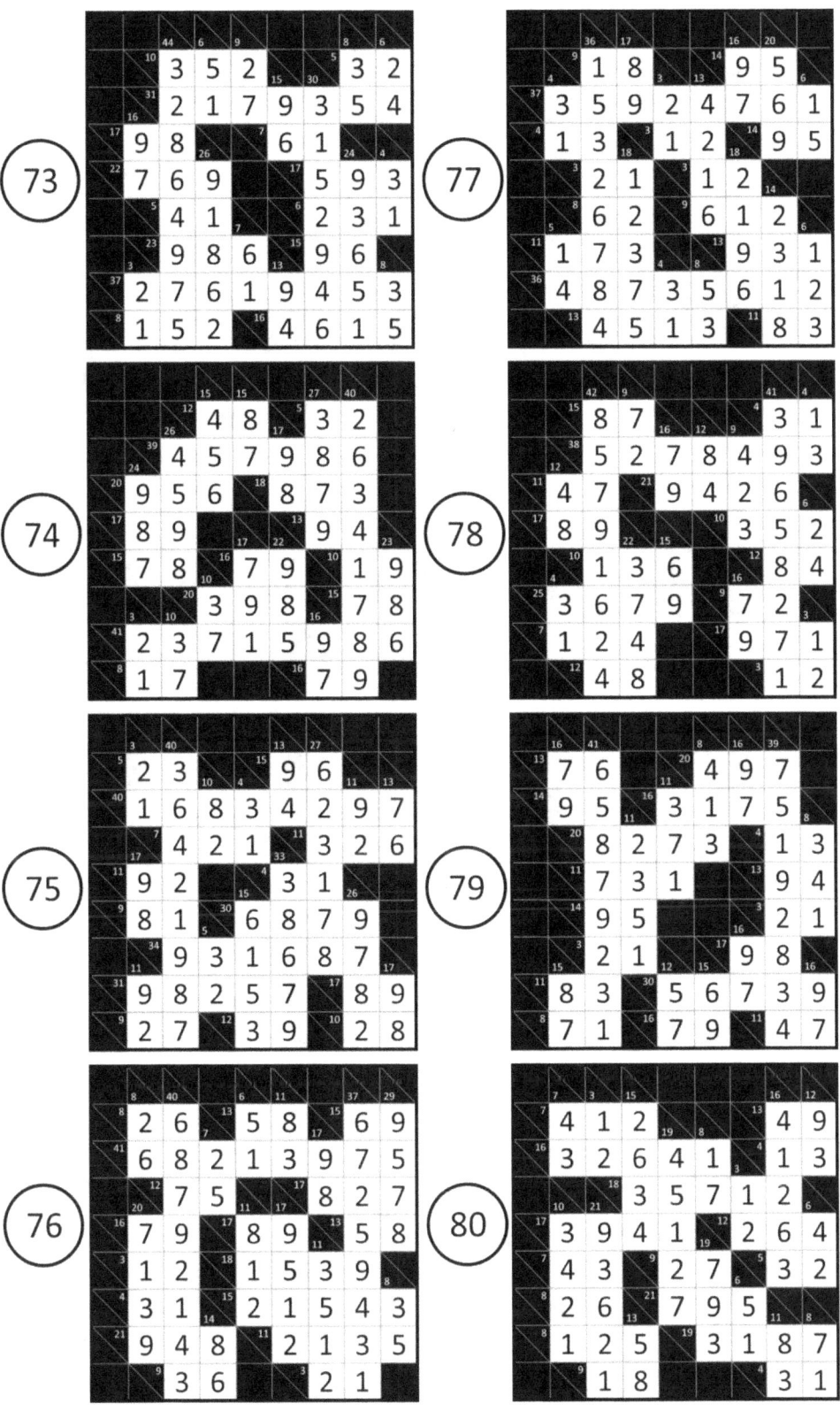

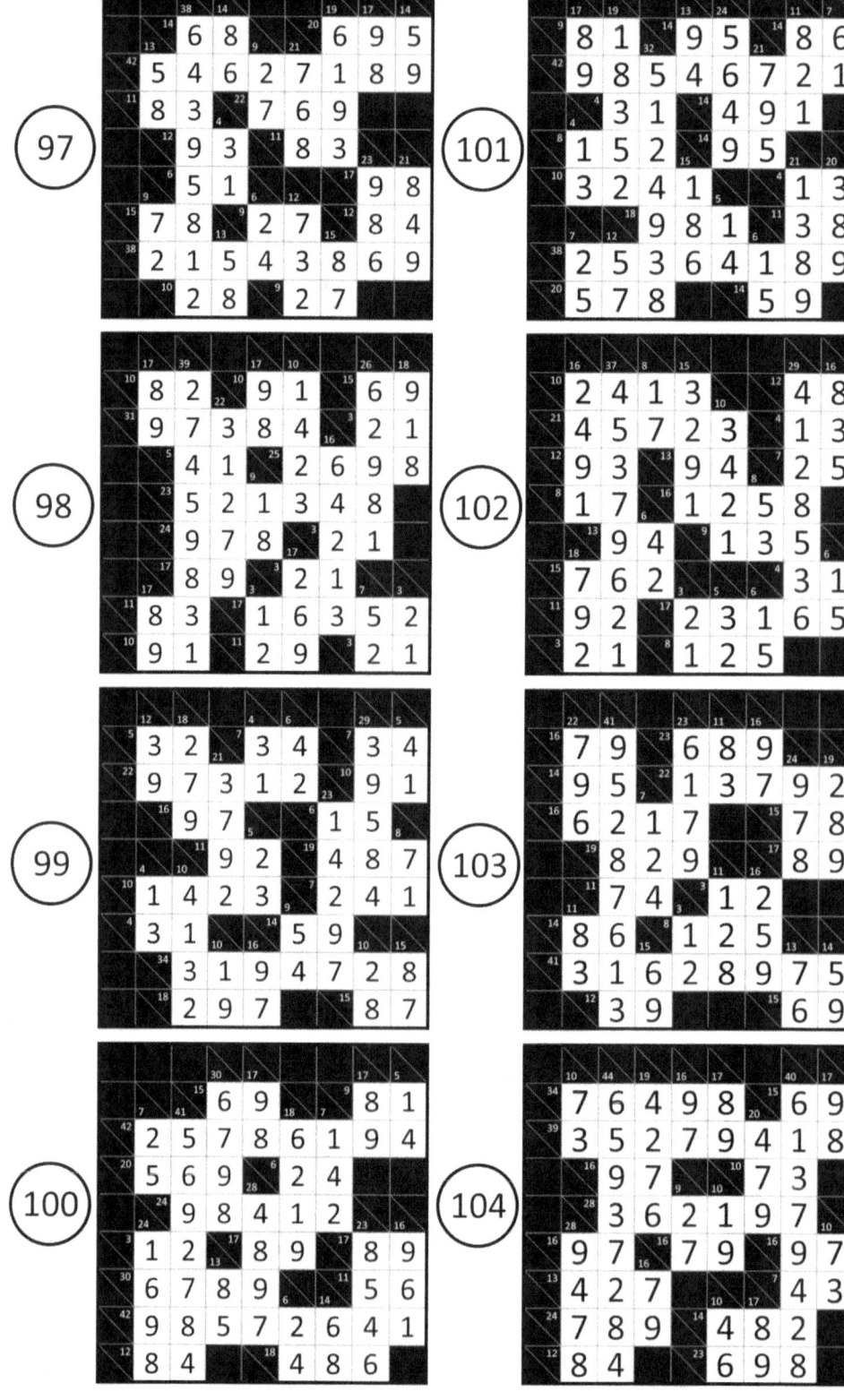

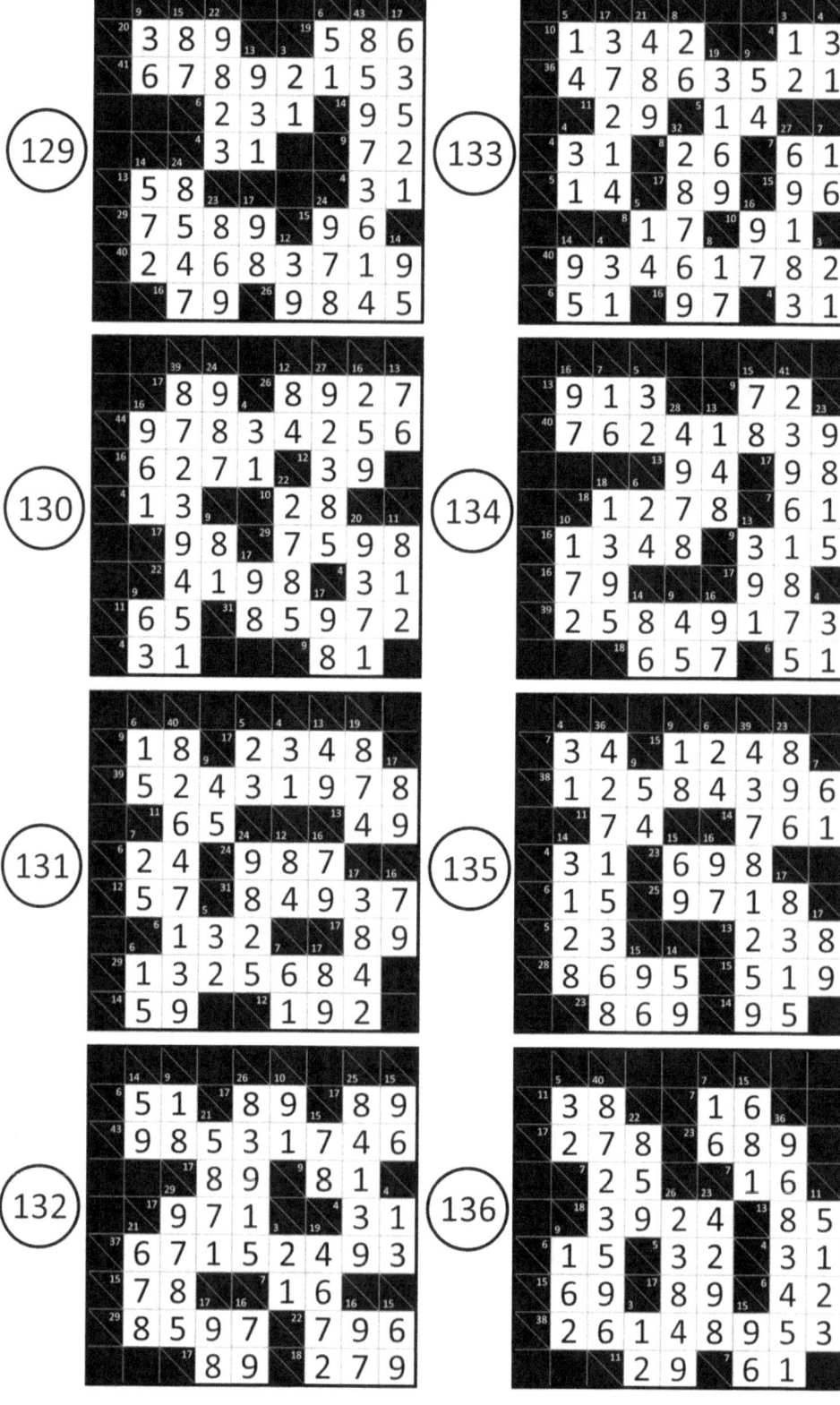

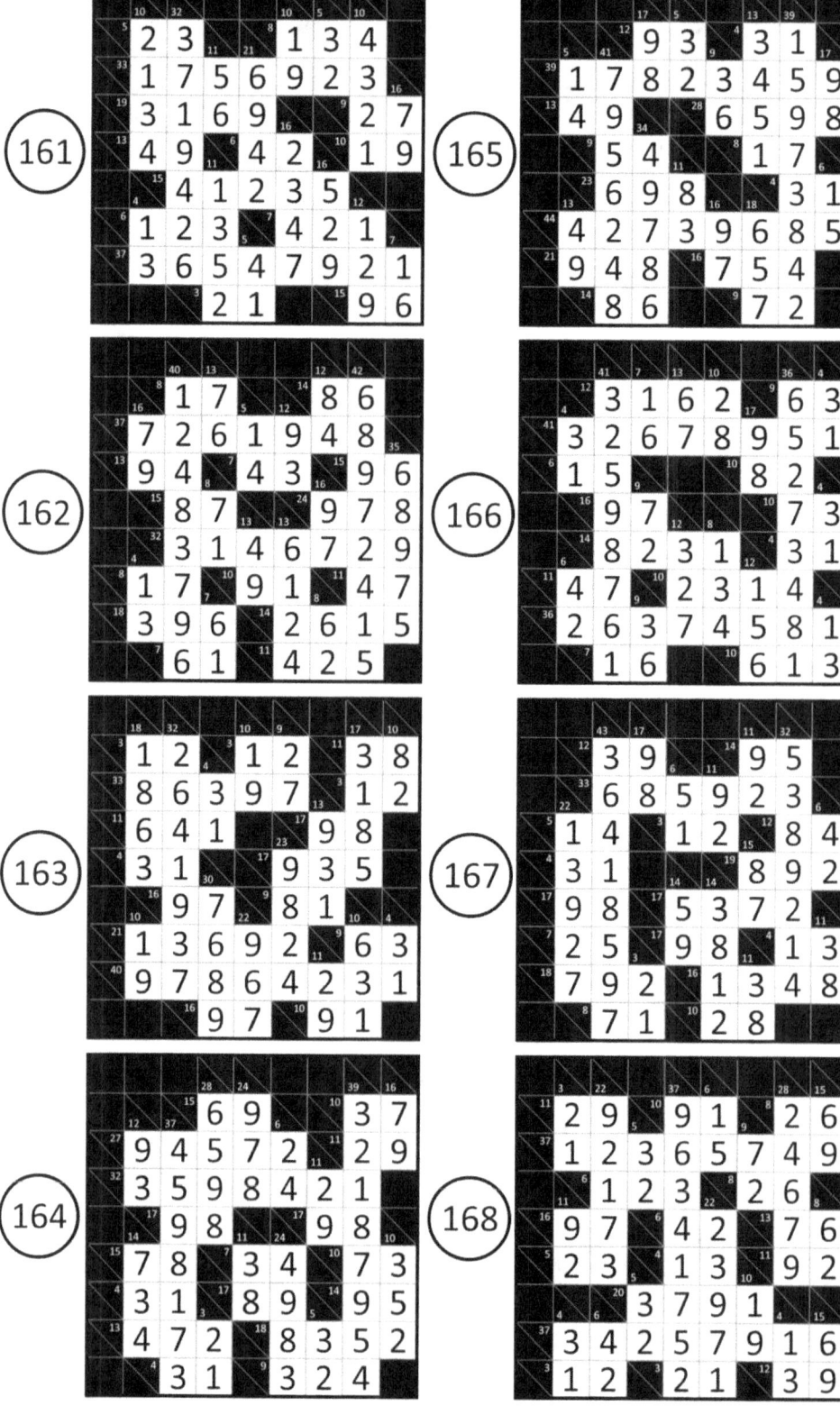

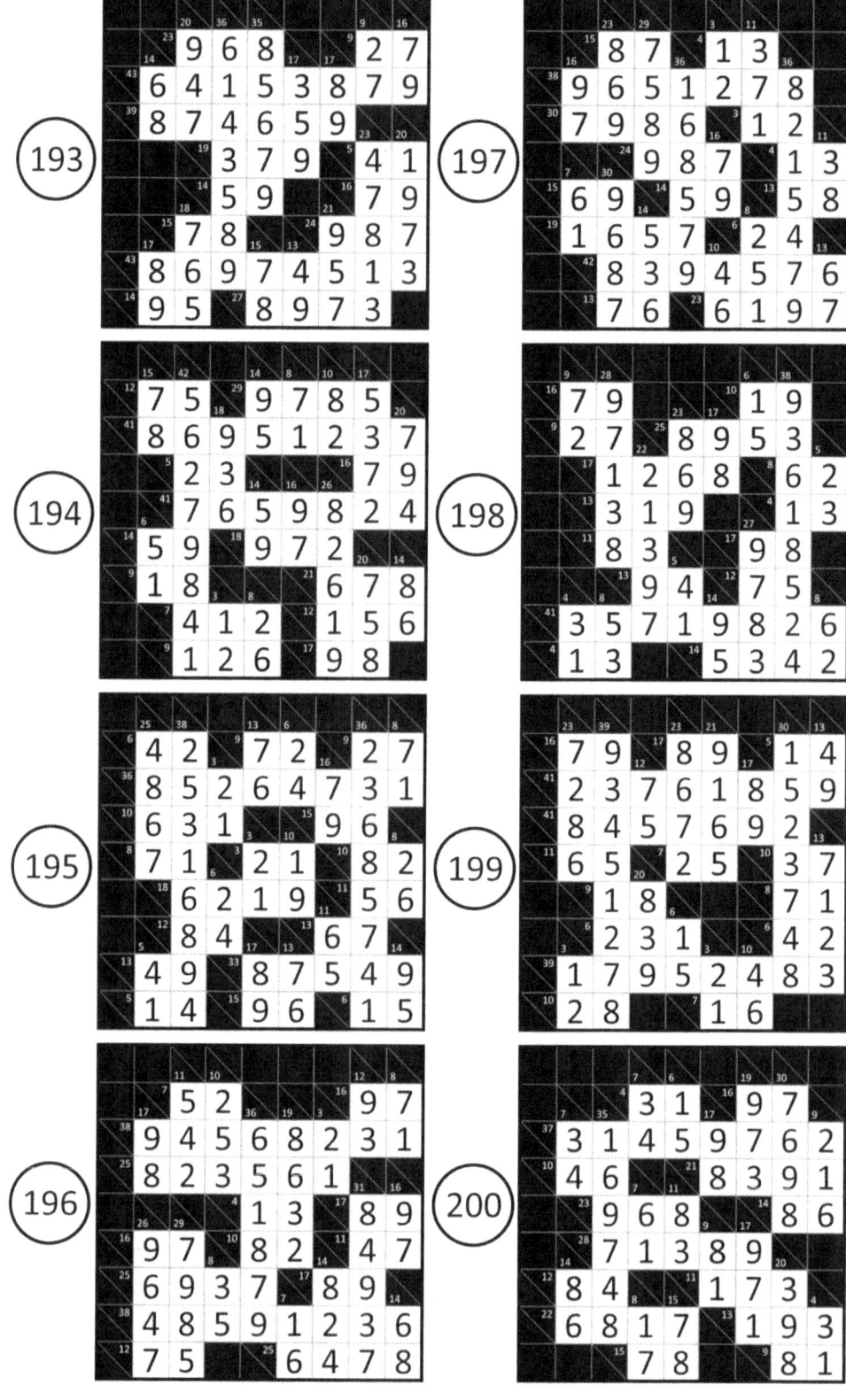

www.ingramcontent.com/pod-product-compliance
Lightning Source LLC
Chambersburg PA
CBHW071058240526
45471CB00016B/2078